Memoir 28

A Color Illustrated Guide To

Constituents, Textures, Cements, and Porosities of Sandstones and Associated Rocks

Peter A. Scholle

U.S. Geological Survey

Published by
The American Association of Petroleum Geologists
with the support of
The American Association of Petroleum Geologists Foundation
Tulsa, Oklahoma, U.S.A., 1979

Published March 1979
Second Printing, March 1981
Library of Congress Catalog Card No. 78-78335
ISBN: 0-89181-304-7

Printed by
Rodgers Litho
Tulsa, Oklahoma

Table of Contents

Introduction

This book is designed as a companion volume to AAPG Memoir 27. As with its predecessor volume, the purpose of this book is to provide identified illustrations of important grains, textures, cements, and porosity types for geologists who may not be specialists in the petrography of sandstones and associated sedimentary rocks.

Sandstone petrography is of particular interest to the explorationist for several reasons. First, it can provide valuable information on the detailed composition of sedimentary rocks. From this, one can often draw conclusions about the lithology, climate, and tectonic history of the source area, as well as predicting the response of such units to a variety of subsurface diagenetic environments. Second, one can acquire significant data on the grain size, sorting, and rounding of sedimentary grains. For lithified sediments this may be the only way to obtain such data, which may be useful in determinations of the transport mechanisms and depositional environment of the sediment. Third, information may be obtained on the post-depositional alteration history of sedimentary rocks. This may include data on compaction, cementation, leaching, fracturing, porosity types, and other factors. These are essential for a proper understanding of reservoir rocks and, commonly, petrography provides the only technique for gathering accurate data on such diagenetic factors.

This book is intended as an introduction for exploration geologists or students and is by no means a complete textbook or treatise. However, it does include a wide variety of color photographs of terrigenous clastic grains, cements, and textures of sandstones and common accessory rock types. Although most of the illustrations are of features seen with the petrographic microscope, some scanning electron micrographs are included. The illustrations were made from samples having as wide a range of lithologies, geologic ages, and localities as possible to insure a fairly representative presentation. In addition, the photographs were generally selected to show the most common grain and textural types encountered by the geologist and to present typical, rather than spectacular, examples of most features. Thus, the book should have applicability to any sandstone petrographic study.

This volume focuses on the descriptive aspects of petrography and includes no text other than figure captions. Bibliographies are provided in each section of the book. For more detailed descriptive and interpretive information, the references listed in both the general and specific bibliographies should be consulted.

The major emphasis of this book is on the four major fabric elements of sandstones: framework grains; detrital fine-grained matrix; cements; and pore space. The dominant framework grains of most sandstones are quartz, feldspars, and rock fragments and these are illustrated in detail. Accessory minerals, such as calcite, mica, glauconite, zircon, magnetite, or organic carbon, are present in most rocks and occasionally are significant rock-forming elements. These minerals are shown in the section entitled "Other detrital grains." Detrital matrix minerals of sandstones and shales are included in a separate section. Framework and matrix grain types can be distinguished readily with the petrographic microscope, especially when supplemented with staining, scanning electron microscopy (SEM), X-ray diffraction, and cathodoluminescence (discussed in a section entitled "Techniques").

The proper identification of framework grains is the key to the classification of sandstones. Examples are provided of the major sandstone types as seen in thin section, and brief summaries are given of the major classification schemes. It is best to consult the original reference, however, before using any particular one because of the complexity of most classifications. In addition, different authors have different definitions of their major compositional end members. For example, some authors include chert with the quartz group while others place chert with the rock fragments.

Rock textures, particularly grain size, shape, sorting, and orientation, are commonly described from thin section, especially when rocks are strongly indurated. Several examples are shown that cover the spectrum from texturally immature to texturally mature rocks. Charts are provided for the estimation of degree of sorting and the description of grain shapes.

Information about the cementation of sandstones is perhaps the most complex, yet potentially valuable, data that can be collected from thin section analysis. This book illustrates the most common cement types in sandstones as well as some complex sequences of multiple generations of cement formation. In addition, other diagenetic fabrics such as replacement, dissolution, compaction, and grain deformation are shown, as are both primary and secondary porosity types.

The information that can be gained from the types of analyses illustrated in this book can be extremely valuable in understanding hydrocarbon reservoir rocks. Such petrographic studies are best undertaken in conjunction with well-log analysis, seismic interpretation, regional geologic synthesis, and other techniques. When applied by the exploration geologist within this larger frame of reference, the results can be remarkably useful. And in the context of overall hydrocarbon exploration costs, the time and expense of such studies are minimal.

Despite the fact that sandstone petrography is a complex and varied subject, it is hoped that, using this color photographic guide, even investigators with little formal petrographic training will be able to examine thin sections under the microscope and interpret the main rock constituents and their diagenetic history. It is hoped that this volume will both stimulate and facilitate petrographic studies of subsurface and outcrop units.

Explanation of Captions

Each photograph in this manual has a description in standard format. The first two lines give the stratigraphic unit, depth (for subsurface samples), and state or country of origin. Lithologic unit names shown in full quotation marks are informal designations; names shown in single quotation marks are ones not recognized by the USGS. The locality data are followed by a description of the photograph. The last line of the caption defines the type of lighting used and the scale of the photograph. The following code is used for lighting:

XN —crossed nicols (crossed polarizers)
PXN —partly crossed nicols
RL —reflected light
RTL —reflected and transmitted light
CL —cathodoluminescence
SEM —scanning electron microscopy
TEM —transmission electron microscopy.

The absence of any symbol indicates that transmitted light with uncrossed polarizers was used. All scales are given as a certain number of millimeters (or micrometers, μm, for scanning electron micrographs) equivalent to a uniform length (1.25 cm or 0.5 in.) for all photographs. Thus, a figure of 0.38 mm indicates that a length of 1.25 cm on the printed picture is equivalent to 0.38 mm on the original specimen.

Acknowledgements

An earlier, shorter version of this book was compiled at the Cities Service Oil Co. Exploration and Production Research Lab, and much of that original version is included here. I would like to express my thanks to Cities Service for their kind permission to publish this work. During the past seven years, many people have contributed samples or photographs to this book. I would particularly like to thank A. S. Boggs, G. A. Desborough, S. P. Dutton, E. J. Dwornik, R. L. Folk, A. J. Gude, III, P. L. Hansley, D. E. Hoyt, C. W. Keighin, R. R. Larson, R. G. Loucks, E. F. McBride, E. D. McKee, C. W. Naeser, E. D. Pittman, Edwin Roedder, V. Schmidt, B. C. Schreiber, R. A. Sheppard, R. F. Sippel, G. W. Smith, C. W. Spencer, R. B. Tripp, T. R. Walker, and C. R. Williamson for the invaluable material they contributed. Finally, I would like to express my appreciation to Sigrid Asher Bolinder, R. L. Folk, T. D. Fouch, P. L. Hansley, C. W. Keighin, E. F. McBride, K. A. Schwab, and C. W. Spencer, who reviewed various drafts of this book and made many constructive suggestions.

35 mm Slide Set

A set of 100 selected photographs from this book are available in the format of 35 mm slides. The set is designed to aid instructors who might use the text for teaching purposes. Photographs included in the slide set are indicated by a ★ next to the scale in the photograph description. The slide set can be ordered from AAPG, P.O. Box 979, Tulsa, OK 74101; cost $128.00 (price subject to change without notice). A similar set of slides for carbonate rock keyed to AAPG *Memoir* 27 "A Color Illustrated Guide to Carbonate Rock Constituents, Textures, Cements and Porosities" is also available for $128.00 from AAPG.

Grain Size Scales for Sediments

The grade scale most commonly used for sediments is the Wentworth scale (actually first proposed by Udden), which is a logarithmic scale in that each grade limit is twice as large as the next smaller grade limit. For more detailed work, sieves have been constructed at intervals $2\sqrt{2}$ and $4\sqrt{2}$.

The ϕ (phi) scale, devised by Krumbein, is a much more convenient way of presenting data than if the values are expressed in millimeters, and is used almost entirely in recent work.

U.S. Standard Sieve Mesh #		Millimeters	Microns (μm)	Phi (φ)	Wentworth Size Class	
		4096		-12		
		1024		-10	Boulder (-8 to -12φ)	GRAVEL
Use		256		- 8		
wire		64		- 6	Cobble (-6 to -8φ)	
squares		16		- 4	Pebble (-2 to -6φ)	
5		4		- 2		
6		3.36		- 1.75		
7		2.83		- 1.5	Granule	
8		2.38		- 1.25		
10		2.00		- 1.0		
12		1.68		- 0.75		
14		1.41		- 0.5	Very coarse sand	
16		1.19		- 0.25		
18		1.00		0.0		
20		0.84		0.25		
25		0.71		0.5	Coarse sand	
30		0.59		0.75		
35	1/2	0.50	500	1.0		SAND
40		0.42	420	1.25		
45		0.35	350	1.5	Medium sand	
50		0.30	300	1.75		
60	1/4	0.25	250	2.0		
70		0.210	210	2.25		
80		0.177	177	2.5	Fine sand	
100		0.149	149	2.75		
120	1/8	0.125	125	3.0		
140		0.105	105	3.25		
170		0.088	88	3.5	Very fine sand	
200		0.074	74	3.75		
230	1/16	0.0625	62.5	4.0		
270		0.053	53	4.25		
325		0.044	44	4.5	Coarse silt	
		0.037	37	4.75		
	1/32	0.031	31	5.0		
Analyzed	1/64	0.0156	15.6	6.0	Medium silt	MUD
by	1/128	0.0078	7.8	7.0	Fine silt	
	1/256	0.0039	3.9	8.0	Very fine silt	
Pipette		0.0020	2.0	9.0		
		0.00098	0.98	10.0	Clay	
or		0.00049	0.49	11.0		
Hydrometer		0.00024	0.24	12.0		
		0.00012	0.12	13.0		
		0.00006	0.06	14.0		

(From Folk, 1968)

Grain
Scales: cont.

Comparison Chart For Visual Percentage
Estimation (After Terry and Chilingar, 1955).

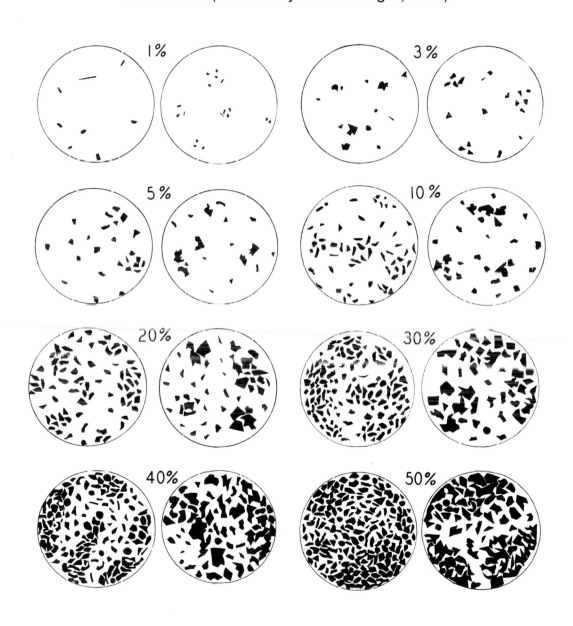

Michel-Lévy Color Chart

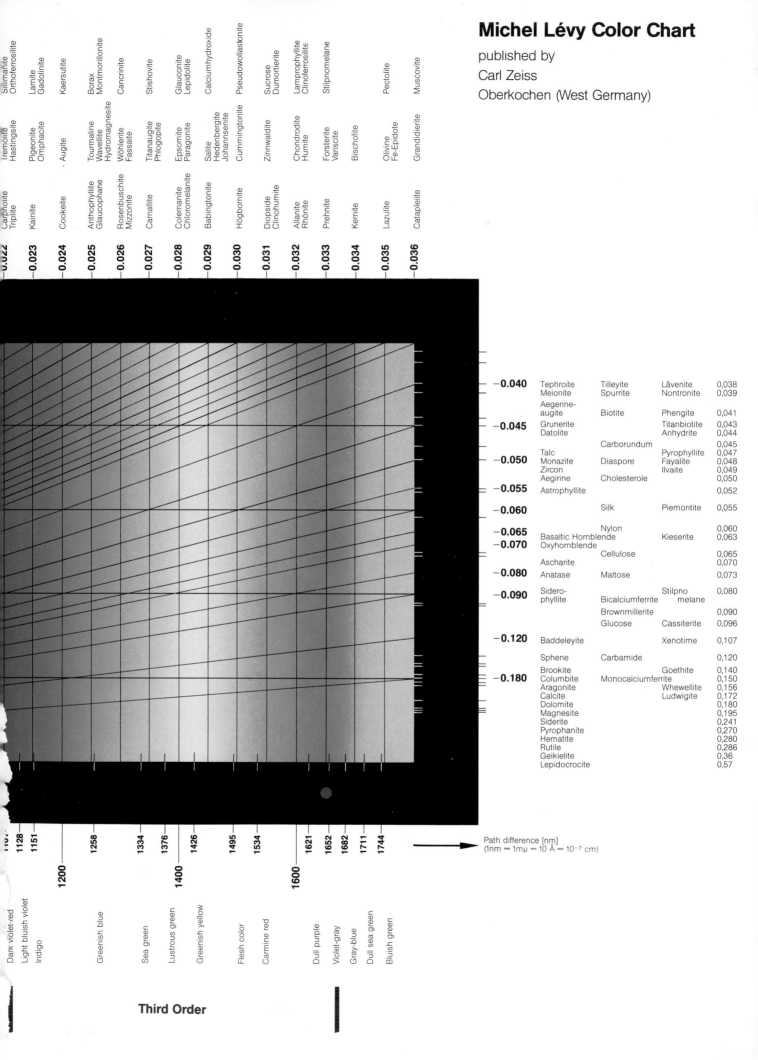

Michel Lévy Color Chart

published by
Carl Zeiss
Oberkochen (West Germany)

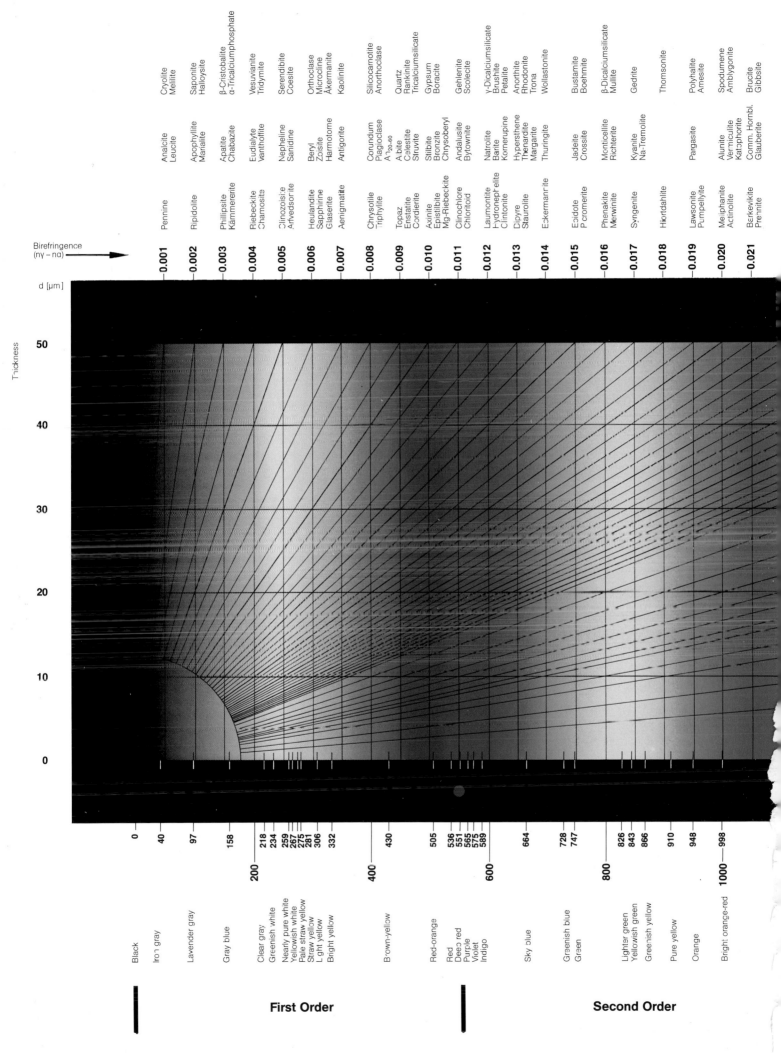

Birefringence (nγ – nα) →

Cryolite			
Melilite			
Analcite			
Leucite			
Pennine	**0.001**		

d [μm]

Thickness

First Order   **Second Order**

Michel-Lévy interference colour chart (birefringence vs. thickness).

Birefringence values across top: 0.001 Cryolite/Melilite/Analcite/Leucite/Pennine, 0.002 Saponite/Halloysite/Apophyllite/Marialite/Ripidolite, 0.003 β-Cristobalite/α-Tricalciumphosphate/Apatite/Chabazite/Phillipsite/Kämmererite, 0.004 Vesuvianite/Tridymite/Eudialyte/Vanthoffite/Riebeckite/Chamosite, 0.005 Serendibite/Coesite/Nepheline/Sanidine/Clinozoisite/Arfvedsonite, 0.006 Orthoclase/Microcline/Åkermanite/Heulandite/Sapphirine/Glaserite, 0.007 Kaolinite/Aenigmatite, 0.008 Silicocarnotite/Anorthoclase/Corundum/Plagioclase/A₆₀-₆₀/Chrysotile/Triphylite, 0.009 Quartz/Rankinite/Tricalciumsilicate/Albite/Celestite/Struvite/Topaz/Enstatite/Cordierite, 0.010 Gypsum/Boracite/Stilbite/Bronzite/Chrysoberyl/Axinite/Epistilbite/Mg-Riebeckite, 0.011 Gehlenite/Scolecite/Andalusite/Bytownite/Clinochlore/Chloritoid, 0.012 γ-Dicalciumsilicate/Brushite/Petalite/Natrolite/Barite/Kornerupine/Laumontite/Hydronephelite/Clintonite, 0.013 Anorthite/Rhodonite/Trona/Hypersthene/Thenardite/Margarite/Dipyre/Staurolite, 0.014 Wollastonite/Thuringite/Eckermanite, 0.015 Bustamite/Boehmite/Jadeite/Crossite/Epidote/Picromerite, 0.016 β-Dicalciumsilicate/Mullite/Monticellite/Richterite/Phenakite/Merwinite, 0.017 Gedrite/Kyanite/Na-Tremolite/Syngenite, 0.018 Thomsonite/Hiortdahlite, 0.019 Polyhalite/Amesite/Pargasite/Lawsonite/Pumpellyite, 0.020 Spodumene/Amblygonite/Alunite/Vermiculite/Katophorite/Meliphanite/Actinolite, 0.021 Brucite/Gibbsite/Comm. Hornbl./Glauberite/Barkevikite/Prehnite.

Path difference scale (bottom, in nm): 0, 40, 97, 158, 200, 218, 234, 259, 267, 275, 281, 306, 332, 400, 430, 505, 536, 551, 565, 575, 589, 600, 664, 728, 747, 800, 826, 843, 866, 910, 948, 998 (1000).

Interference colours (bottom): Black, Iron gray, Lavender gray, Gray blue, Clear gray, Greenish white, Nearly pure white, Yellowish white, Pale straw yellow, Straw yellow, Light yellow, Bright yellow, Brown-yellow, Red-orange, Red, Deep red, Purple, Violet, Indigo, Sky blue, Greenish blue, Green, Lighter green, Yellowish green, Greenish yellow, Pure yellow, Orange, Bright orange-red.

Thickness scale (left, in μm): 0, 10, 20, 30, 40, 50.

General

Berry, L. G., and B. Mason, 1959, Mineralogy: San Francisco, W. H. Freeman and Co., 630 p.

Blatt, Harvey, G. V. Middleton, and R. C. Murray, 1972, Origin of sedimentary rocks. Englewood Cliffs, New Jersey, Prentice-Hall Inc., 634 p.

Bloss, F. D., 1961, An introduction to the methods of optical crystallography: New York, Holt, Rinehart and Winston, 294 p.

Boswell, P. G. H., 1933, On the mineralogy of sedimentary rocks: London, Thomas Murby and Co., 393 p.

Carozzi, A. V., 1960, Microscopic sedimentary petrography: New York, John Wiley and Sons, 485 p.

Carver, R. E., ed., 1971, Procedures in sedimentary petrology: New York, Wiley-Interscience, 672 p.

Correns, C. W., 1969, Introduction to mineralogy: New York, Springer-Verlag, 484 p.

Deer, W. A., R. A. Howie, and J. Zussman, 1962, Rock forming minerals, Volume 1, Ortho- and Ringsilicates: New York, John Wiley and Sons, 333 p.

—— 1963a, Rock forming minerals, Volume 2, Chain silicates: New York, John Wiley and Sons, 379 p.

—— 1963b, Rock forming minerals, Volume 3, Framework silicates: New York, John Wiley and Sons, 435 p.

Durrell, Cordell, 1949, A key to the common rock-forming minerals in thin section: San Francisco, W. H. Freeman and Co., 16 p.

El-Hinnawi, E. E., 1966, Methods in chemical and mineral microscopy: New York, Elsevier, 222 p.

Folk, R. L., 1974, Petrology of sedimentary rocks: Austin, Texas, Hemphill's Book Store, 182 p.

Füchtbauer, Hans, 1974a, Sediments and sedimentary rocks, 1, part II: New York, Halsted Press, 464 p.

—— 1974b, Some problems of diagenesis in sandstones, in Sedimentation argilo-sableuse et diagenese: Centre Rech. Pau Bull., v. 8, p. 391-403.

Füchtbauer, Hans, and German Müller, 1977, Sedimente und Sedimentgesteine (Sediment-Petrologie, Teil 2): Stuttgart, E. Schweitzerbart'sche Verlagsbuchhandlung, 783 p.

Krumbein, W. C., and F. J. Pettijohn, 1961, Manual of sedimentary petrography: New York, Appleton-Century-Crofts, 549 p.

Larsen, E. S., and H. Berman, 1934, The microscopic determination of the nonopaque minerals, 2nd ed.: U.S. Geological Survey Bull. 848, 266 p.

Loughnan, F. C., 1969, Chemical weathering of the silicate minerals: New York, Elsevier, 154 p.

Milner, H. B., 1962a, Sedimentary petrography, Volume 1, Methods in sedimentary petrography, 4th ed.: London, George Allen and Unwin Ltd., 643 p.

—— 1962b, Sedimentary petrography, Volume II, Principles and applications, 4th ed.: London, George Allen and Unwin Ltd., 715 p.

Müller, G., 1967, Methods in sedimentary petrology: New York, Hafner, 283 p.

Pettijohn, F. J., 1975, Sedimentary rocks, 3rd ed.: New York, Harper and Row, 628 p.

—— P. E. Potter, and R. Siever, 1972, Sand and sandstone: New York, Springer-Verlag, 618 p.

Saggerson, E. P., 1975, Identification tables for minerals in thin sections: London and New York, Longman, 378 p.

Solomon, M., and R. Green, 1966, A chart for designing modal analysis by point counting: Geol. Rundschau, v. 55, p. 844-848.

General Selected Bibliography

Torry, R. D., and G. V. Chilingar, 1955, Summary of "Concerning some additional aids in studying sedimentary formations," by M. S. Shvetsov: Jour. Sed. Petrology, v. 25, p. 229-234.

Tickell, F. G., 1965, The techniques of sedimentary mineralogy: Developments in sedimentology, Volume 4: Amsterdam, Elsevier, 220 p.

Van Der Plas, L., and A. C. Tobi, 1965, A chart for judging the reliability of point counting results: Am. Jour. Sci., v. 263, p. 87-90.

Vanders, Iris, and P. F. Kerr, 1967, Mineral recognition: New York, Wiley-Interscience, 316 p.

Wahlstrom, E. E., 1955, Petrographic mineralogy: New York, John Wiley and Sons, 408 p.

Williams, Howel, F. J. Turner, and C. M. Gilbert, 1954, Petrography: San Francisco, W. H. Freeman and Co., 406 p.

Winchell, A. N., and H. Winchell, 1951, Elements of optical mineralogy, Part II, Descriptions of minerals: New York, John Wiley and Sons, 551 p.

Winkler, H. G., 1974, Petrogenesis of metamorphic rocks, 3rd ed.: New York, Springer-Verlag, 320 p.

General Geochemical

Berner, R. A., 1971, Principles of chemical sedimentology: New York, McGraw-Hill, 240 p.

Degens, E. T., 1965, Geochemistry of sediments. A brief survey: Englewood Cliffs, New Jersey, Prentice-Hall Inc., 342 p.

Ernst, Werner, 1970, Geochemical facies analysis: Methods in geochemistry and geophysics, Volume 11: New York, Elsevier, 152 p.

Garrels, R. M., 1960, Mineral equilibria: New York, Harper and Bros., 254 p.

—— and C. L. Christ, 1965, Solutions, minerals, and equilibria: New York, Harper and Row, 450 p.

—— and F. T. Mackenzie, 1971, Evolution of sedimentary rocks: New York, W. W. Norton and Co., 397 p.

Krauskopf, K. B., 1967, Introduction to geochemistry: New York, McGraw-Hill, 721 p.

Pettijohn, F. J., 1963, Chemical composition of sandstones—excluding carbonate and volcanic sands, *in* Data of geochemistry, 6th ed.: U.S. Geological Survey Prof. Paper 440-S, 21 p.

Robie, R. A., and D. R. Waldbaum, 1968, Thermodynamic properties of minerals and related substances at 298.15°K (25.0°C) and one atmosphere (1.013 bars) pressure and at higher temperatures: U.S. Geological Survey Bull. 1259, 256 p.

General Applied

Cotera, A. S., 1956, Petrology of the Cretaceous Woodbine Sand in northeast Texas: Master's thesis, University of Texas.

Davies, D. K., and R. R. Berg, 1969, Sedimentary characteristics of Muddy barrier-bar reservoir and lagoonal trap at Bell Creek field, *in* Eastern Montana Symposium: Montana Geol. Soc. 20th Ann. Conf., p. 97-105.

—— and F. G. Ethridge, 1975, Sandstone composition and depositional environment: AAPG Bull., v. 59, p. 239-264.

Folk, R. L., 1960, Petrography and origin of the Tuscarora, Rose Hill, and Keefer Formations, Lower and Middle Silurian, of eastern West Virginia: Jour. Sed. Petrology, v. 30, p. 1-58.

—— 1962, Petrography and origin of the Silurian Rochester and McKenzie Shales, Morgan County, West Virginia: Jour. Sed. Petrology, v. 32, p. 539-578.

Heald, M. T., and G. F. Baker, 1977, Diagenesis of the Mt. Simon and Rose Run sandstones in western West Virginia and southern Ohio: Jour. Sed. Petrology, v. 47, p. 66-77.

Krynine, P. D., 1950, Petrology, stratigraphy, and origin of the Triassic sedimentary rocks of Connecticut: Conn. Geol. and Nat. History Survey Bull., no. 73, 247 p.

Merino, Enrique, 1975, Diagenesis in Tertiary sandstone from Kettleman North Dome, California. I. Diagenetic mineralogy: Jour. Sed. Petrology, v. 45, p. 320-336.

Pryor, W. A., 1971, Petrology of the Weissliegendes Sandstone in the Harz and Werra-Fulda areas, Germany: Geol. Rundschau, v. 60, p. 524-552.

Sabins, F. F., Jr., 1963, Anatomy of stratigraphic trap, Bisti field, New Mexico: AAPG Bull., v. 47, p. 193-228.

Schwab, F. L., 1975, Framework mineralogy and chemical composition of continental margin-type sandstone: Geology, v. 3, p. 487-490.

Todd, T. W., and R. L. Folk, 1957, Basal Claiborne of Texas, record of Appalachian tectonism during Eocene: AAPG Bull., v. 41, p. 2545-2566.

Detrital Grains

Quartz

Upper Cambrian Gatesburg Fm. ★
Pennsylvania

Quartz, seen here, is the dominant framework grain of most sandstones and is
recognized primarily by its low birefringence, lack of cleavage and twinning, low
positive relief, and uniaxial positive interference figure. Many quartz types have
been defined, based mainly on the number
of subcrystals in the grain and their extinction behavior when the stage is rotated
under crossed polarizers. Some of these
quartz types may be related to specific
source rocks.

XN   0.15 mm

Pleistocene Yellowstone Group (tuff)
Wyoming

A single-crystal, inclusion-poor, uniform or
straight-extinction quartz grain with well
developed crystal outlines surrounded by a
dark matrix of fused glass shards with
pronounced flow texture. Straight-extinction grains extinguish completely uniform-
ly when the stage is rotated (less than 1
degree of undulosity). The combination of
well developed crystal outlines and straight
extinction is common in quartz of volcanic
origin.

0.27 mm

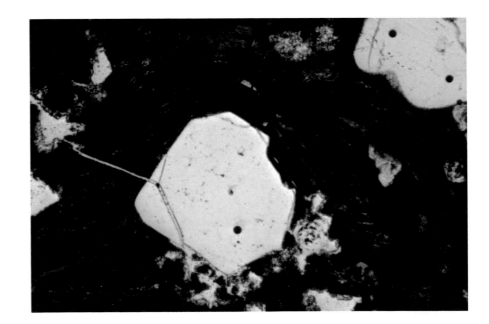

Pleistocene Yellowstone Group (tuff) ★
Wyoming

Detail of volcanic quartz crystal. This grain
has straight extinction, a euhedral outline,
and a large "negative crystal" or vacuole.
The vacuole has the same crystallographic
orientation as the complete quartz grain,
hence the term "negative crystal". This
feature is common but not ubiquitous in
quartz of volcanic origin.

XN   0.24 mm

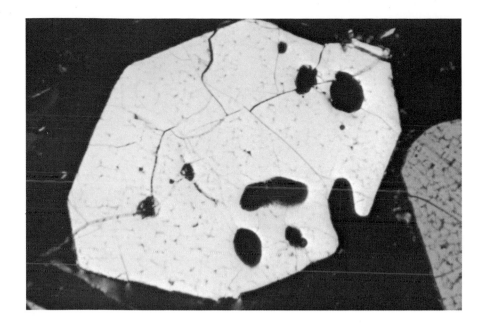

Pleistocene Yellowstone Group (tuff)
Wyoming

A volcanic (β) quartz grain with euhedral, bipyramidal outline. Euhedral shape, embayments, straight extinction, and scarcity of inclusions are all indicative of an extrusive igneous source, but none, by itself, is conclusive evidence.

XN   0.38 mm

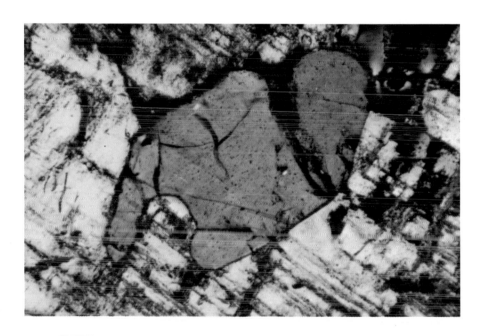

Precambrian Llano Series (granite)
Texas

A nondetrital quartz grain (in a nonsedimentary "source" rock) showing rounded outline and embayment. Thus, not all original grains are angular, and embayment is not restricted to volcanic quartz. Quartz crystal (photo center) is surrounded by plagioclase feldspar.

XN   0.24 mm

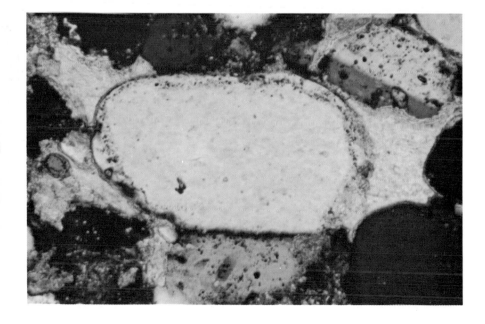

Tertiary 'Vieja Group'
Texas

A quartz grain of volcanic origin. Grain shows a zone of inclusions near its outer margins and a second zone with fewer inclusions at the outside of the grain. This zonation, similar in appearance to abraded authigenic quartz overgrowths, is a common feature in grains from volcanic sources.

XN   0.15 mm

Lower Permian Brushy Canyon Fm. ★
Texas

An authigenic overgrowth on a detrital quartz grain. The well rounded nucleus is outlined by a thin layer of inclusions (probably clay and iron oxides) on its surface. The authigenic overgrowth shows euhedral crystal shape where fully developed. Such euhedral outlines produced by overgrowth must be carefully distinguished from volcanic quartz zonation.

XN   0.06 mm

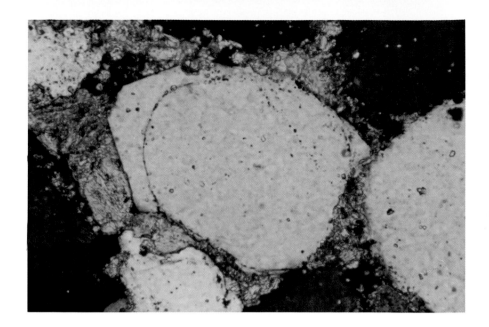

Tertiary Horse Springs Fm.
Nevada

Abundant volcanic glass shards composed of opalline silica are seen here in a largely unaltered and undeformed state. Because opal is an unstable mineral under near surface conditions, such shards are frequently devitrified, replaced, or dissolved entirely. However, when preserved, they are excellent indicators of a volcanic provenance. Further examples are shown in the section on "Associated sediments".

0.10 mm

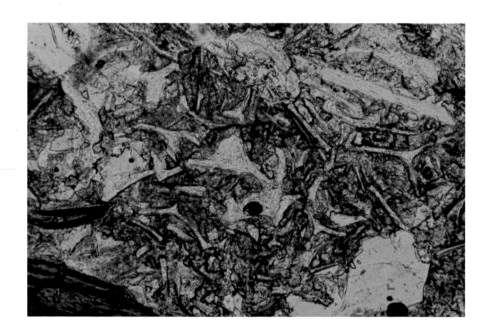

Tertiary 'Vieja Group'
Texas

A single-crystal, straight extinction, detrital quartz grain ("common quartz" or "normal igneous quartz" of Krynine, 1940). Grain extinguishes completely under crossed polarizers with less than 1 degree of stage rotation. Such quartz is supplied by many types of source rocks, and may be selectively concentrated during weathering and transportation.

XN   0.08 mm

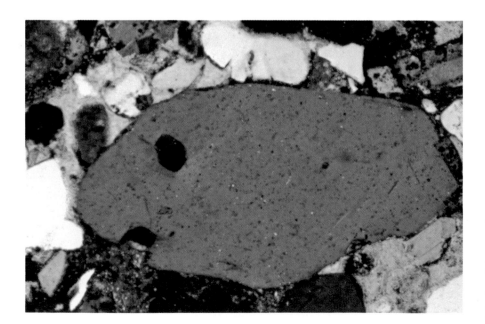

Upper Cambrian Gatesburg Fm.
Pennsylvania

Large grain in center is a single-crystal, slightly undulose quartz grain ("end phase" or "igneous" quartz of Krynine, 1940 and 1946). Grain extinguishes completely with between 1 and 5 degrees of stage rotation. Such extinction behavior is best studied using a universal stage but can be done with less accuracy on a flat stage. Slightly undulose quartz can be derived from most types of source terrains.

XN   0.15 mm

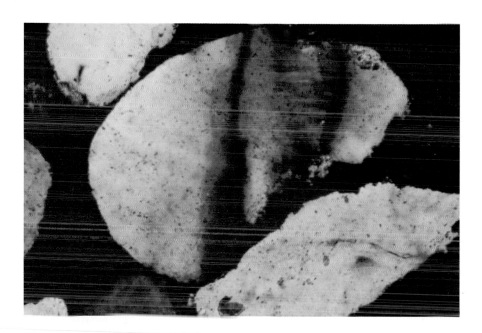

Middle Silurian Clinton Fm. ★
Virginia

A single-crystal quartz grain with strongly undulose extinction. Grain extinguishes completely with more than 5 degrees of stage rotation. Such grains may be more abundant in strained source rocks (especially metamorphics), but evidence is still incomplete.

XN   0.06 mm

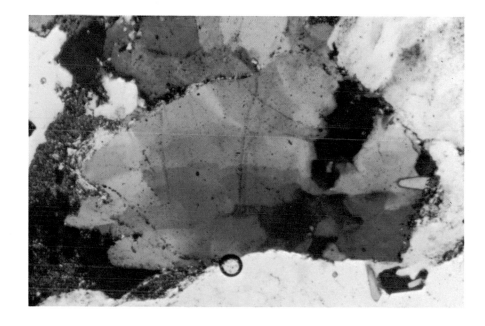

Upper Triassic New Haven Arkose
Connecticut

A semicomposite quartz grain with slightly undulose extinction. Grain consists of a number of separate quartz crystals with very closely aligned optic c-axes. Such grains are common in hydrothermal veins but also occur in many other metamorphic and plutonic rock types.

XN   0.15 mm

Upper Triassic New Haven Arkose ★
Connecticut

A detrital grain of composite or polycrystalline quartz (the "recrystallized metamorphic" quartz type of Krynine, 1940).
C-axes of individual subcrystals show considerable variation in orientation; crystal
boundaries are straight and crystal shapes
are somewhat elongate. This grain is
probably of high grade metamorphic
origin. Many sandstone classifications
would include this as a quartz grain but
some would class this as a rock fragment.

XN   0.15 mm

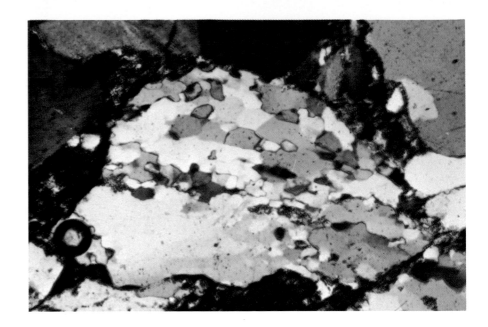

Ordovician(?) kyanite schist
Connecticut

Composite or polycrystalline quartz in a
high grade metamorphic source rock. Note
relatively straight boundaries between
equant grains, as well as straight to slightly
undulose extinction. This quartz type is a
fairly good indicator of a metamorphic
source when 10 or more subcrystals are
present in a detrital grain.

XN   0.38 mm

Precambrian "Adirondack gneiss"
New York

Polycrystalline or composite quartz in a
high grade metamorphic source rock. Note
irregular, crennulate boundaries between
strongly elongate grains (equivalent to
Krynine's "schistose" or "pressure"
quartz). The presence of 10 or more individual crystals in a single detrital grain, is
an excellent indicator of a metamorphic
source.

XN   0.24 mm

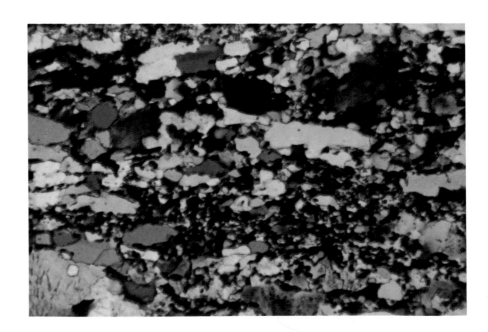

Jurassic Curtis Fm.
Utah

A complex detrital quartz grain. Although this is a composite quartz, the component crystals are themselves semicomposite. The crystals also have sutured contacts, a preferred crystallographic fabric, and only slight elongation. This grain is probably derived from a metamorphic terrain.

XN   0.27 mm

Upper Cretaceous Kogosukruk Tongue of
  Prince Creek Fm. *
Alaska

Detrital chert fragments are seen as speckled grains composed of microcrystalline quartz. These grains are derived from a sedimentary source. Chert is either classed as a quartz type or as a sedimentary rock fragment depending on which sandstone classification is used. Chert is discussed more fully in the section on "Associated sediments."

XN   0.15 mm

Middle Ordovician Lander Ss. Mbr. of
  Bighorn Dolomite
Wyoming

Quartz crystals with few inclusions of either vacuole or microlite variety. This is the most common type of quartz and is found in nearly all source rocks. The unusually high birefringence colors of quartz in this photo are due to the fact that the thin section is approximately 5 μm thicker than the standard 30 μm.

XN   0.10 mm

8

Devonian(?) gneiss
Massachusetts

A quartz grain with a large epidote micro
lite (mineral inclusion) and semicomposite
extinction. Such mineral inclusions may be
useful both for correlation purposes and
for interpretation of source area composition.

XN   0.38 mm

Cambrian(?) Harrison Summit Quartzite
Idaho

Quartz grains with several muscovite microlites and slightly undulose extinction.
Because muscovite can occur in a wide
range of source rocks, these particular
inclusions have relatively little provenance
significance. However, such inclusions can
still be useful as stratigraphic markers in
correlation.

XN   0.06 mm

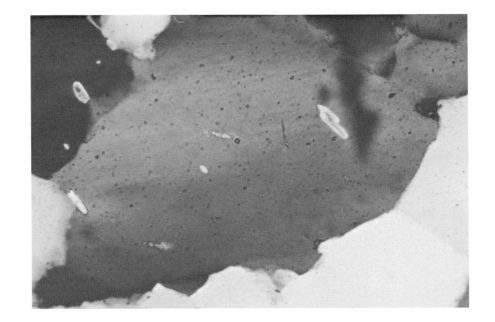

Paleozoic andalusite schist ⋆
New Hampshire

Quartz grains with abundant needle-shaped
mineral inclusions. The inclusions in this
case are sillimanite, but actinolite, tremo-
lite, rutile, and other minerals can also be
found as needle-like inclusions in quartz.
Detrital quartz grains with sillimanite
inclusions are excellent evidence for a
metamorphic source area.

XN   0.06 mm

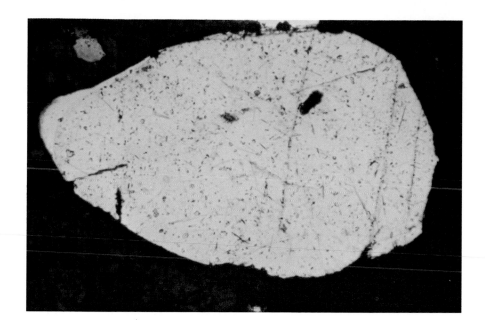

Lower Pennsylvanian Sharon Mbr. of
  Pottsville Fm.
Ohio

A detrital quartz grain with numerous very
small vacuoles and needle-shaped mineral
inclusions. Needles here are mainly rutile.
The quartz grain has some overgrowths
which are much poorer in inclusions, and
is surrounded by a later hematite cement.

XN   0.08 mm

Cretaceous Travis Peak Fm. *
Texas

A quartz crystal with very abundant
vacuoles (probably liquid filled) and semicomposite extinction (Krynine's "hydrothermal" quartz). Such vacuole rich quartz
does appear to be most commonly derived
from hydrothermal-vein sources.

XN   0.24 mm

Lower Cretaceous Patula Arkose *
Mexico

A detrital quartz grain with warped, sub-
parallel lines of very small bubbles which
have been termed Boehm lamellae. They
are the product of intense strain deformation of quartz grains. One must be careful
to distinguish between detrital strained
grains and *in situ* deformation. Boehm
lamellae can also be confused with some
types of feldspar twinning.

XN   0.24 mm

10

Devonian Cairn Fm.
Canada (Alberta)

Authigenic quartz crystals in a limestone.
These crystals form by replacement,
usually of carbonate rocks. A detrital
nucleus is commonly present in the crystal
centers, and these nuclei, in combination
with abundant carbonate inclusions and
euhedral crystal outlines, are used to
identify authigenic sedimentary quartz.

XN   0.22 mm

Devonian Cairn Fm.
Canada (Alberta)

Detail of an authigenic quartz crystal replacement of limestone matrix. Note
abundance of carbonate inclusions (as
well as calcite crystals which are partly
overlapping atop the quartz). Authigenic
replacement quartz almost always contains
inclusions of the original host sediment.

XN   0.05 mm

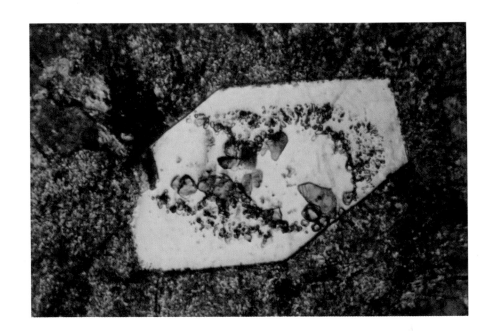

Holocene grus on Mesozoic granite
California

Quartz grains, as seen here, are normally
recognizable in the scanning electron
microscope on the basis of their conchoidal
fracture and lack of cleavage. Photo by
D. E. Hoyt.

SEM   500 µm

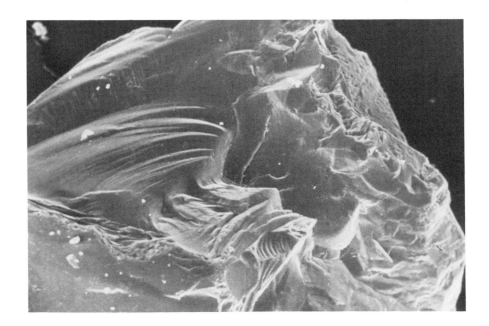

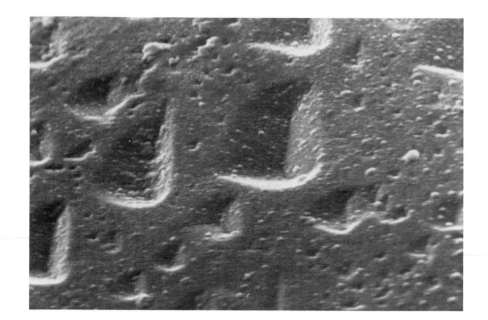

Quaternary grus on Precambrian
  Packsaddle Schist
Texas

Quartz grains have a wide variety of surface textures which can be seen at high magnification, particularly in SEM. V-shaped indentations are seen in this example. Various features have been interpreted as being indicative of particular source rocks, transport processes, depositional environments, or diagenetic settings. Photo by D. E. Hoyt.

SEM   0.7 μm

Upper Permian Bell Canyon Fm.
Texas 1,388 m (4,580 ft)

Quartz which has formed diagenetic overgrowths is also readily recognizable in SEM. The overgrowths here have smooth, partially interlocking crystal faces. In this example, quartz overgrowths and chlorite cement are intergrown. Photo by C. R. Williamson.

SEM   10 μm

Selected Detrital Grain Bibliography

Quartz

Adams, S. F., 1920, A microscopic study of vein quartz: Econ. Geology, v. 15, p. 623-664.

Andersen, D. W., and M. D. Picard, 1971, Quartz extinction in siltstone: Geol. Soc. America Bull., v. 82, p. 181-186.

Basu, Abhijit, 1975, Re-evaluation of the use of undulatory extinction and polycrystallinity in detrital quartz for provenance interpretation: Jour. Sed. Petrology, v. 45, p. 873-882.

Blatt, H., 1959, Effect of size and genetic quartz type on sphericity and form of beach sediments, northern New Jersey: Jour. Sed. Petrology, v. 29, p. 197-206.

——— 1967a, Original characteristics of clastic quartz grains: Jour. Sed. Petrology, v. 37, p. 401-424.

——— 1967b, Provenance determinations and recycling of sediments: Jour. Sed. Petrology, v. 37, p. 1031-1044.

——— and J. M. Christie, 1963, Undulatory extinction in quartz of igneous and metamorphic rocks and its significance in provenance studies of sedimentary rocks: Jour. Sed. Petrology, v. 33, p. 559-579.

Bloss, F. D., 1957, Anisotropy of fracture in quartz: Am. Jour. Sci., v. 255, p. 214-225.

Bokman, J., 1952, Clastic quartz particles as indices of provenance: Jour. Sed. Petrology, v. 22, p. 17-24.

Chaklader, A. C. D., 1963, Deformation of quartz crystals at the transformation temperature: Nature, v. 197, p. 791-792.

Cleary, W. J., and J. R. Conolly, 1971, Distribution and genesis of quartz in a piedmont-coastal plain environment: Geol. Soc. America Bull., v. 82, p. 2755-2766.

——— 1972, Embayed quartz grains in soils and their significance: Jour. Sed. Petrology, v. 42, p. 899-904.

Conolly, J. R., 1965, The occurrence of polycrystallinity and undulatory extinction in quartz in sandstones: Jour. Sed. Petrology, v. 35, p. 116-135.

Crook, K. A. W., 1968, Weathering and roundness of quartz sand grains: Sedimentology, v. 11, p. 171-182.

Dennen, W. H., 1967, Trace elements in quartz as indicators of provenance: Geol. Soc. America Bull., v. 78, p. 125-130.

Ethridge, F. G., and D. K. Davies, 1969, Quartz as an indicator of depositional environment: an example from the Eocene (Claiborne) of Texas: Geol. Soc. America Abs. with Programs, v. 1, pt. 2, p. 11.

Greensmith, J. T., 1963, Clastic quartz, provenance and sedimentation: Nature, v. 197, p. 345-347.

Harrell, James, and Harvey Blatt, 1978, Polycrystallinity: effect on the durability of detrital quartz: Jour. Sed. Petrology, v. 48, p. 25-30.

Keller, W. D., and R. F. Littlefield, 1950, Inclusions in the quartz of igneous and metamorphic rocks: Jour. Sed. Petrology, v. 20, p. 74-84.

Krauskopf, K. B., 1959, The geochemistry of silica in sedimentary environments, in H. A. Ireland, ed., Silica in sediments: Soc. Econ. Paleontologists and Mineralogists Spec. Pub. No. 7, p. 4-19.

Krinsley, D. H., and I. J. Smalley, 1973, Shape and nature of small sedimentary quartz particles: Science, v. 180, p. 1277-1279.

Krynine, P. D., 1940, Petrology and genesis of the Third Bradford Sand: Penn. State College Mineral Industries Expt. Sta. Bull. 29, 134 p.

——— 1946, Microscopic morphology of quartz types: 2nd Cong. Panam. Ing. Min. Geol. Anales, v. 3, p. 35-49.

Kuenen, P. H., 1969, Origin of quartz silt: Jour. Sed. Petrology, v. 39, p. 1631-1633.

Morey, G. W., R. O. Fournier, and J. J. Rowe, 1962, The solubility of quartz in water in the temperature interval from 25°C to 300°C: Geochim. et Cosmochim. Acta, v. 26, p. 1029-1044.

Moss, A. J., 1966, Origin, shaping, and significance of quartz sand grains: Jour. Geol. Soc. Australia, v. 13, p. 97-136.

Renton, J. J., M. T. Heald, and C. B. Cecil, 1969, Experimental investigation of pressure solution of quartz: Jour. Sed. Petrology, v. 39, p. 1107-1117.

Siever, R., 1957, The silica budget in the sedimentary cycle: Am. Mineralogist, v. 42, p. 821-841.

Sosman, R. B., 1927, The properties of silica: Am. Chem. Soc. Monograph Series, New York, Reinhold Publishing Co., 856 p.

van Lier, J. A., P. L. DeBruyn, and J. T. Overbeck, 1960, The solubility of quartz: Jour. Phys. Chem., v. 64, p. 1675-1682.

Wood, D. G., and H. Blatt, 1968, Influence of roundness and internal structure on the chemical stability of quartz grains: Geol. Soc. America Program with Abstracts, Mexico City, p. 327-328.

Young, S. W., 1976, Petrographic texture of detrital polycrystalline quartz as an aid to interpreting crystalline source rocks: Jour. Sed. Petrology, v. 46, p. 595-603.

Quartz Surface Texture

Baker, H. W., Jr., 1976, Environmental sensitivity of submicroscopic surface textures on quartz sand grains—a statistical evaluation: Jour. Sed. Petrology, v. 46, p. 871-880.

Brown, J. E., 1973, Depositional histories of sand grains from surface textures: Nature, v. 242, p. 396-398.

Coch, N. K., and D. H. Krinsley, 1971, Comparison of stratigraphic and electron microscopic studies in Virginia Pleistocene coastal sediments: Jour. Geology, v. 79, p. 426-437.

Doornkamp, J. C., and D. H. Krinsley, 1971, Electron microscopy applied to quartz grains from a tropical environment: Sedimentology, v. 17, p. 89-101.

Hodgson, A. V., and W. B. Scott, 1970, The identification of ancient beach sands by the combination of size analysis and electron microscopy: Sedimentology, v. 14, p. 67-75.

Ingersoll, R. V., 1974, Surface textures of first cycle quartz sand grains: Jour. Sed. Petrology, v. 44, p. 151-157.

Krinsley, D. H., and J. C. Doornkamp, 1973, Atlas of quartz sand surface textures: Cambridge, England, Cambridge Earth Sci. Series, 91 p.

——— and L. Cavallero, 1970, Scanning electron microscopic examination of periglacial eolian sands from Long Island, New York: Jour. Sed. Petrology, v. 40, p. 1345-1350.

——— and J. Donahue, 1968, Diagenetic surface textures on quartz grains in limestones: Jour. Sed. Petrology, v. 38, p. 859-862.

——— and J. Donahue, 1968, Environmental interpretation of sand grain surface textures by electron microscopy: Geol. Soc. America Bull., v. 79, p. 743-748.

—— and T. Takahashi, 1962, Surface textures of sand grains: an application of electron microscopy: Science, v. 135, p. 923-925.

Kuenen, P. H., and W. G. Perdok, 1961, Frosting of quartz grains: Kon. Ned. Akad. v. Wetenschappen, Amsterdam, Proc., Ser. B, v. 64, p. 343-345.

—— 1962, Experimental abrasion 5. Frosting and defrosting of quartz grains: Jour. Geology, v. 70, p. 648-658.

Margolis, S. V., 1968, Electron microscopy of chemical solution and mechanical abrasion features on quartz sand grains: Sed. Geology, v. 2, p. 243-256.

—— and D. H. Krinsley, 1971, Submicroscopic frosting on eolian and subaqueous quartz sand grains: Geol. Soc. America Bull., v. 82, p. 3395-3406.

—— 1974, Processes of formation and environmental occurrence of microfeatures on detrital quartz grains: Am. Jour. Sci., v. 274, p. 449-464.

Porter, J. J., 1962, Electron microscopy of sand surface texture: Jour. Soc. Petrology, v. 32, p. 124-135.

Schneider, H. E., 1970, Problems of quartz grain morphoscopy: Sedimentology, v. 14, p. 325-335.

Scholle, P. A., and D. E. Hoyt, 1973, Quartz grain surface textures from various source rocks: Geol. Soc. America Abs. with Programs, v. 6, p. 797-798.

Setlow, L. W., and R. P. Karpovich, 1972, "Glacial" micro-textures on quartz and heavy mineral sand grains from the littoral environment: Jour. Sed. Petrology, v. 42, p. 864-875.

Subramanian, V., 1975, Origin of surface pits on quartz as revealed by scanning electron microscopy: Jour. Sed. Petrology, v. 45, p. 530-534.

Walker, T. R., 1957, Frosting of quartz grains by carbonate replacement: Geol. Soc. America Bull., v. 68, p. 267-268.

Whalley, W. B., and D. H. Krinsley, 1974, A scanning electron microscope study of surface textures of quartz grains from glacial environments: Sedimentology, v. 21, p. 87-105.

Wolf, M. J., 1967, An electron microscope study of the surface texture of sand grains from a basal conglomerate: Sedimentology, v. 8, p. 239-247.

Feldspars

16

Ordovician Newtown Gneiss ★
Connecticut

Complex twinning in a plagioclase feldspar grain. Exact types of twinning are best determined on a universal stage, but albite, carlsbad, and pericline twins are probably present here. Twin types can sometimes indicate source area. Pink tint in grains resulted from staining for plagioclase.

XN   0.38 mm

Tertiary Horse Spring Fm. ★
Nevada

Plagioclase feldspars (unstained) in a volcanic sandstone. Note euhedral crystal outlines, well-defined crystal zoning (growth-composition lines) and the albite twinning. All these features, taken together, are indicative of volcanic plagioclase.

XN   0.38 mm

Ordovician Newtown Gneiss ★
Connecticut

A complex plagioclase grain. Core and outer rim are clearly of different compositions. Twinning extends throughout grain, but core shows considerable alteration (mainly vacuolization and sericitization) while rim is largely unaltered. Illustrates that alteration can take place in source rocks and is highly composition dependent.

XN   0.10 mm

Ordovician Newtown Gneiss *
Connecticut

Microcline feldspar with typical microcline
grid twinning. Although such twinning is
characteristic of most triclinic alkali
feldspars, it is most commonly shown by
microcline. Some small inclusions of
plagioclase with albite twinning are present here.

XN   0.24 mm

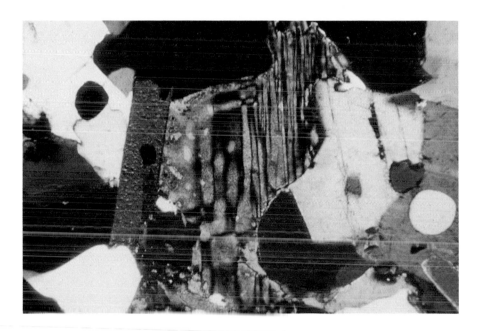

Precambrian Hitchcock Lake Mbr. of
    Waterbury Gneiss
Connecticut

Yellowish grain in center is a microcline
feldspar with spindle twinning—the irregular lamellar twins can often be used to
distinguish microcline. Yellow color is a
stain for K-spar. The brown, elongate grain
directly to the left of the microcline is
biotite.

XN   0.30 mm

Tertiary intrusive *
Nevada

Sanidine, a potassium feldspar, is shown
here with no twinning but stained yellow
using a sodium cobaltinitrite solution.
Staining is often the fastest and most
reliable method of identifying untwinned
feldspars. Matrix in this sample is volcanic
glass with a pronounced flow texture.

0.28 mm

18

Pennsylvanian-Permian Sangre de Cristo
    Fm. *
New Mexico

Untwinned orthoclase. Many feldspars, of
all types, are untwinned and, when unstained, can be distinguished only by such
features as alteration (seen here), cleavage,
fracture, relief, or optic figures. The
alteration seen here is coarser, more extensive, and more regularly oriented than in
quartz grains. Orthoclase is characterized
by two cleavages, low relief (below quartz),
and a biaxial negative figure (2V of 60 to
85°). Orthoclase may show Carlsbad twinning.

XN   0.08 mm

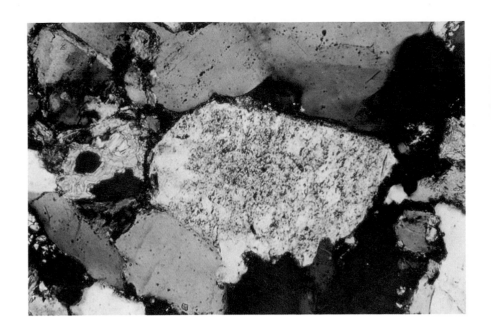

Lower Cretaceous Patula Arkose
Mexico

The plagioclase feldspars in this example
can be readily distinguished from quartz
because of the darker appearance of the
feldspars caused by secondary vacuolization and sericitization. It is difficult but
important to distinguish detrital altered
grains from those which have decayed
in situ.

0.30 mm

Lower Cretaceous Patula Arkose
Mexico

Same as previous photo but with crossed
polarizers. High birefringence and speckled,
micaceous texture indicate alteration is
mainly sericitization. Note that some
feldspars are altered whereas others are
largely unaffected. This may reflect compositional differences which yield different
susceptibilities to alteration, or it may
indicate that the alteration took place in
the source area where only some grains
were affected.

XN   0.30 mm

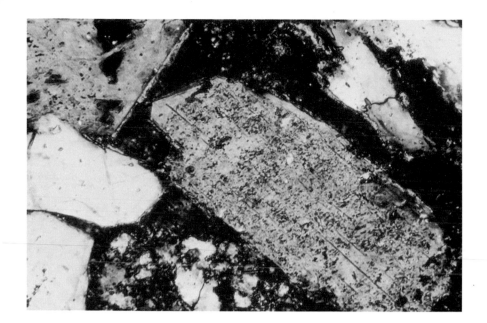

Upper Cretaceous Kogosukruk Tongue of
Prince Creek Fm.
Alaska

Grain in center is an altered feldspar
(probably an orthoclase). Alteration is
primarily by vacuolization. The altered
part of the grain is, however, surrounded
by an unaltered overgrowth. Possibly the
grain was altered in the source area and was
overgrown *in situ*. Other grains include
chert, quartz, and plagioclase.

XN   0.12 mm

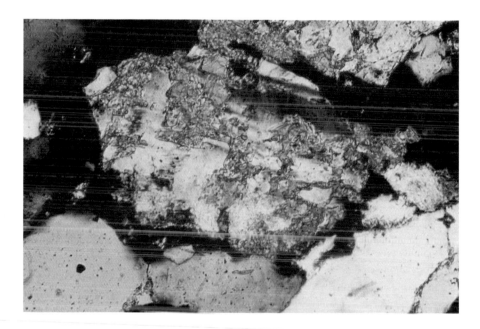

Pennsylvanian-Permian Sangre de Cristo
Fm. *
New Mexico

A plagioclase feldspar largely replaced by
calcite. To accurately determine the
primary composition of sandstones it is
often necessary to recognize feldspars in
very advanced stages of destruction. Calcite
replacement is one very common form of
diagenetic alteration.

XN   0.08 mm

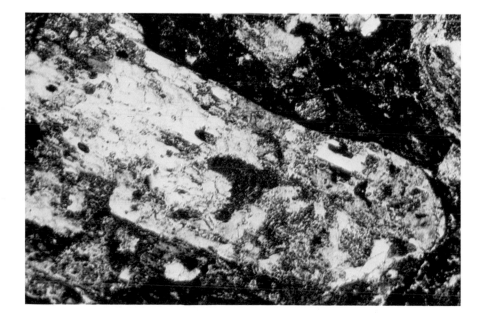

Devonian Old Red Ss.
Northern Ireland

A plagioclase feldspar with multiple types
of alteration. This grain has been partly
replaced by clay minerals as well as calcite.
It also has large etched voids which were
filled with clayey matrix through infiltra-
tion or precipitation.

XN   0.10 mm

Upper Cretaceous Kogosukruk Tongue of
    Prince Creek Fm. *
Alaska

Feldspars commonly undergo leaching in
subsurface environments. This example
shows a relatively early stage in which
there has been only partial removal of the
interior zones of the feldspar. Note the
strong control of dissolution patterns along
prefered crystallographic directions within
the mineral. Green color in photograph is
stained mounting material.

0.06 mm

Oligocene Frio Fm.
Texas 920 m (3,017 ft)

An example of virtually complete subsurface dissolution of a detrital feldspar.
Traces of clays and other impurities mark
former crystallographic planes of the feldspar as well as coating the exterior of the
leached grain. This type of dissolution is
common in feldspars within overpressured
strata of the Gulf Coast. Commonly the
dissolution is really a removal of calcite
which replaced the feldspars. Blue color is
stained impregnating and mounting material. Photo by R. G. Loucks.

0.07 mm

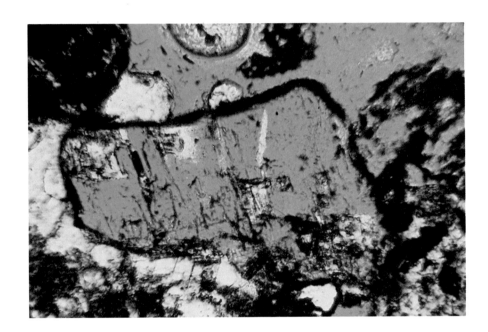

Triassic Shinarump Mbr. of Chinle Fm.
Utah

Grain in center of photo is a feldspar which
is clearly differentiable from the surrounding quartz grains on the basis of cleavage
and inclusions. Note the strong orientation of fractures in the feldspar grain as
opposed to the quartz. Porosity is stained
pale green in this section.

0.24 mm

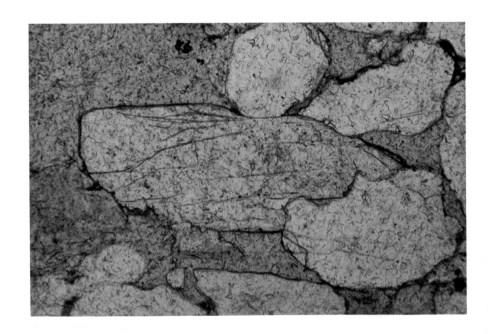

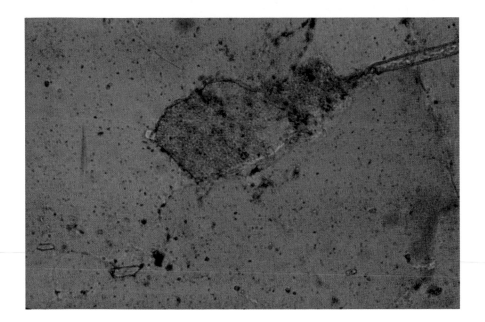

Cambrian(?) Harrison Summit Quartzite
Idaho

An untwinned feldspar (with a slight pink stain for plagioclase), surrounded by quartz grains. Note the slight relief difference between the two minerals as outlined by Becke lines. This can be used for identification of untwinned, unstained feldspars where they are in direct contact with quartz.

0.06 mm

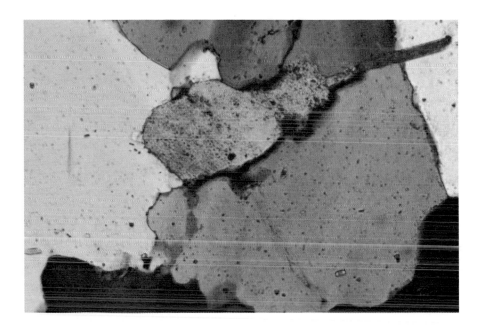

Cambrian(?) Harrison Summit Quartzite
Idaho

Same as previous photo but with crossed polarizers. Illustrates the absence of twinning in this grain and the lack of significant birefringence differences between the feldspar and surrounding quartz crystals.

XN   0.06 mm

Upper Cretaceous Frontier Fm.
Wyoming

A K-feldspar as seen in SEM. Feldspars are generally identifiable in SEM view because of their well developed cleavages, as seen here.

SEM   5 μm

Upper Permian Bell Canyon Fm.
Texas 1,390 m (4,560 ft)

Feldspars which have undergone partial
dissolution are also readily recognizable in
SEM. In this example, the grain has been
deeply etched along preferred crystallographic planes; some subsequent authigenic feldspar growth may also have taken
place. Photo by C. R. Williamson.

SEM   8 μm

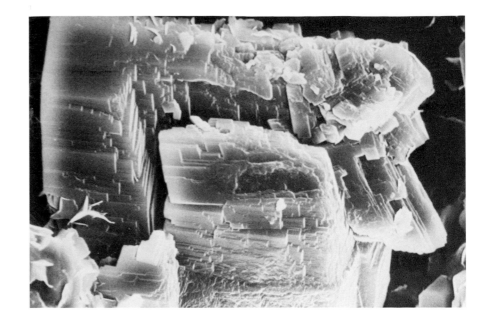

Upper Cretaceous Monte Antola Fm.
Italy

Feldspars can form as authigenic crystals
in clastic terrigenous and calcareous
sediments. In this case, albite is replacing a
limestone matrix. Note crystal outlines,
twinning, and abundance of inclusions.

XN   0.06 mm

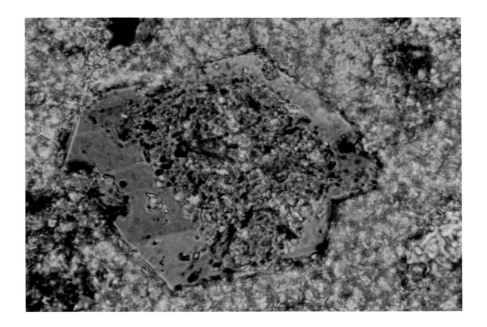

Upper Cretaceous Kogosukruk Tongue of
    Prince Creek Fm.
Alaska

Feldspars with extensive authigenic overgrowths are seen here. These are untwinned
K-feldspars with well developed interlocking overgrowths which could be
mistaken for quartz overgrowths without
careful observation. Porosity is stained
with green epoxy.

                0.15 mm

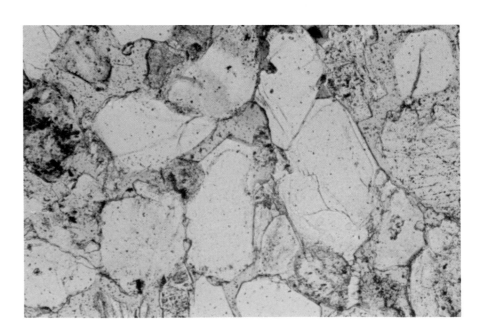

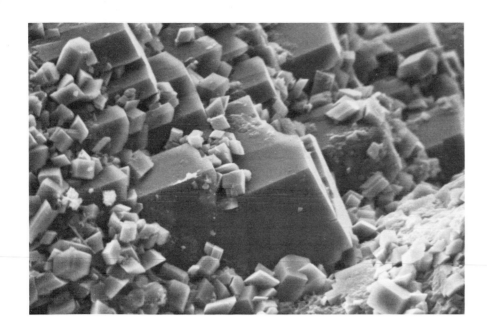

Miocene 'Hayner Ranch Fm.'
New Mexico

Small authigenic K-feldspar crystals are present here in association with a detrital feldspar which has been etched and then authigenically overgrown. Photo by C. W. Keighin, courtesy of T. R. Walker.

SEM   10 μm

Selected Feldspar Bibliography

Bailey, E. H., and R. E. Stevens, 1960, Selective staining of K-feldspar and plagioclase on rock slabs and thin section: Am. Mineralogist, v. 45, p. 1020-1025.

Barth, T. F. W., 1969, Feldspars: New York, Wiley-Interscience, 259 p.

Bostock, H. H., 1966, Feldspar and quartz phenocrysts in the Shingle Creek porphyry, British Columbia: Geol. Survey Canada Bull. 126.

Emmons, R. C., ed., 1953, Selected petrogenic relationships of plagioclase: Geol. Soc. America Mem. 52, 142 p.

Goldsmith, J. R., and F. Laves, 1954, The microcline-sanidine stability relations: Geochim. et Cosmochim. Acta, v. 5, p. 1-9.

Kastner, M., 1970, Feldspars as provenance and physiochemical indicators: Geol. Soc. America Abs. with Programs, v. 2, no. 7, p. 591.

Krynine, P. D., 1942, Provenance versus mineral stability as a controlling factor in the composition of sediments: Geol. Soc. America Bull., v. 53, p. 1850-1851.

Laniz, R. V., R. E. Stevens, and M. B. Norman, 1964, Staining of plagioclase feldspar and other minerals with F. D. and C. Red No. 2: U.S. Geol. Survey Prof. Paper 501-B, p. 152-153.

Laves, F., 1952, Phase relations of the alkali feldspars I. Introductory remarks: Jour. Geology, v. 60, p. 436-450.

Mackie, William, 1899, The feldspars present in sedimentary rocks as indicators of conditions of contemporaneous climate: Trans. Edinburgh Geol. Soc., v. 7, p. 443-468.

Martens, J. H. C., 1931, Persistence of feldspar in beach sands: Am. Mineralogist, v. 16, p. 526-531.

Megaw, H. D., 1959, Order-disorder in the feldspars: Mineral. Mag., v. 32, p. 226-241.

Pittman, E. D., 1970, Plagioclase feldspar as an indicator of provenance in sedimentary rocks: Jour. Sed. Petrology, v. 40, p. 591-598.

Rimsaite, J., 1967, Optical heterogeneity of feldspars observed in diverse Canadian rocks: Schweitz. Min. Petrog. Mitt., v. 47, p. 61-67.

Slemmons, D. B., 1962, Determination of volcanic and plutonic plagioclases using a three- or four-axis universal stage: Geol. Soc. America Spec. Paper 69, 64 p.

Smith, J. V., 1974, Feldspar minerals, 2. Chemical and textural properties: New York, Springer Verlag, 690 p.

Todd, T. W., 1968, Paleoclimatology and the relative stability of feldspar minerals under atmospheric conditions: Jour. Sed. Petrology, v. 38, p. 832-844.

Turner, F. J., 1951, Observations on twinning of plagioclase in metamorphic rocks: Am. Mineralogist, v. 36, p. 581-589.

Van Der Plas, L., 1966, The identification of detrital feldspars: Developments in sedimentology, Volume 6: New York, Elsevier, 305 p.

Wright, T. L., 1968, X-ray and optical study of alkali feldspar: II. An X-ray method for determining the composition and structural state from measurement of 2θ values for three reflections: Am. Mineralogist, v. 53, p. 88-104.

—— and D. B. Stewart, 1968, X-ray and optical study of alkali feldspar: I. Determination of composition and structural state from refined unit-cell parameters and 2V: Am. Mineralogist, v. 53, p. 38-87.

Rock Fragments

Triassic Chinle Fm. *
Arizona

Sedimentary rock fragments (SRF's)—shale
clasts. Shale fragments are difficult to distinguish from low rank metamorphic rock
fragments, especially slates. SRF's are
generally softer and thus more likely to be
deformed or embayed by adjacent harder
grains, as seen in this example. Individual
clay crystal orientation within this rock
fragment is poor, also favoring an SRF
interpretation.

0.10 mm

Triassic Chinle Fm. *
Arizona

Same as previous view but with crossed
polarizers. Note poor orientation and high
birefringence of clay minerals within SRF.
These features indicate that the grain is
probably an illitic shale clast.

XN   0.10 mm

Jurassic Salt Wash Mbr. of Morrison Fm. *
Colorado

Dark brown grain is a detrital clast of shale
or claystone; a SRF. The extremely finely
crystalline nature of the constituent clay
minerals and the association within the
rock with other types of SRF's (not shown
here) help to distinguish the sedimentary
origin of this clast. Note dark-brown rims
of authigenic-clay cement in this sample.

0.10 mm

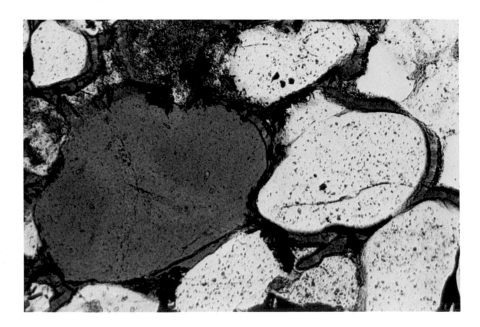

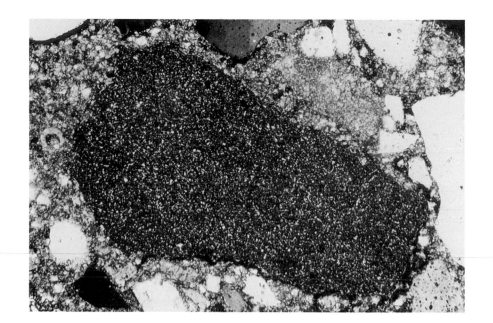

Cretaceous Travis Peak Cgl.
Texas

Sedimentary rock fragment—chert. Note the very uniform microcrystalline quartz with no visible relict texture. Grain is surrounded by carbonate matrix and cement. Chert derived from sedimentary sources can be mistaken for very finely crystalline volcanic rock fragments or clay clasts if not carefully examined.

XN   0.10 mm

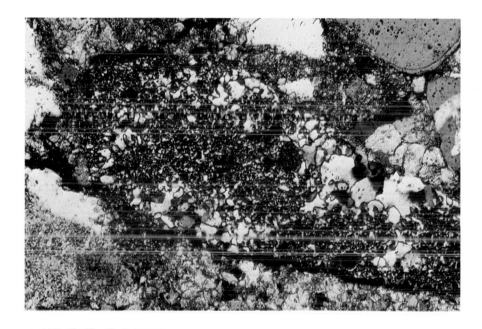

Cretaceous Travis Peak Cgl. ★
Texas

A detrital chert fragment (SRF) with a combination of microquartz and megaquartz. Although most modern sandstone classifications consider detrital chert as a rock fragment, some include it with quartz. More detailed descriptions of chert types are provided in the section on "Associated sediments".

XN   0.10 mm

Jurassic Curtis Fm.
Utah

A detrital chalcedony fragment (SRF). Chalcedony is a microscopically fibrous form of quartz common as a void filling or replacement fabric in sedimentary rocks. Here the chalcedony is surrounded by carbonate cement crystals. As with cherts, a wide variety of chalcedonic textures can be found.

XN   0.27 mm

Permian Abo Ss.
New Mexico

Carbonate rock fragments (SRF's) with a hematite-stained matrix. Polymict calcitic fragments, some with recrystallized fossil fragments, are visible in this rock which would be classed as a calclithite (Folk, 1968). Calcitic SRF's weather and abrade readily and thus are found in abundance only in extremely arid settings or along fault scarps in fanglomeratic deposits which have undergone only short transport.

XN   0.27 mm

Triassic Dockum Gp. *
Texas

Abundant carbonate rock fragments. These grains are less varied in composition and texture than those in the previous example. The uniformity of composition and lack of fossil fragments make it likely that this rock is a reworked caliche (a calcium carbonate-rich soil crust).

XN   0.38 mm

Cretaceous Rieselberger Ss.
Germany

A detrital dolomite fragment (SRF). Note the rhombic shape of constituent crystals as well as their pronounced zoning with cloudy centers and clear rims. These are good criteria for the recognition of dolomite although many dolomites do not show either characteristic. Staining is the most reliable technique for identification of detrital or authigenic dolomite.

XN   0.22 mm

Cretaceous Corwin Fm.
Alaska

Sedimentary rock fragments—dolomite, limestone, and chert clasts. A large dolomitic rock fragment can be seen in upper left; other single dolomite rhombs are also detrital, although this is commonly difficult to prove. Detrital chert fragments and quartz grains are present throughout sample. This material is clearly derived from a sedimentary source.

XN   0.15 mm

Cretaceous Corwin Fm.
Alaska

Abundant isolated dolomite rhombs of probable detrital origin (same rock as previous photo). Lack of replacement relations, rounded corners, consistent equivalence of sediment-grain size and dolomite-crystal size, and presence of associated polycrystalline dolomite rock clasts all can be used as evidence of a detrital origin of dolomite.

XN   0.12 mm

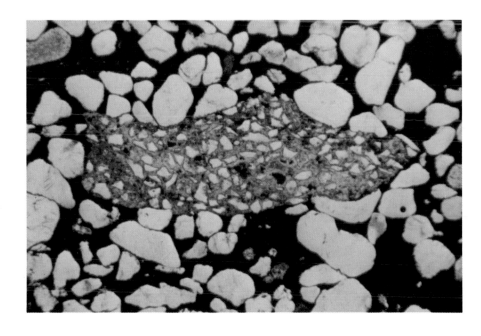

Silurian Rose Hill Fm. ★
Virginia

A detrital siltstone clast (SRF). Siltstone and sandstone clasts are relatively scarce because both rock types tend to break down completely into their component grains. This example shows angular quartz silt grains embedded in a phosphatic matrix. Rock is hematite cemented. Although they are good indicators of a sedimentary source, sandstone and siltstone fragments rarely survive extensive transport.

0.30 mm

Cretaceous Tuktu Fm.
Alaska

A low-rank metamorphic rock fragment (MRF) which has been deformed during compaction. The grain in center is most probably a micaceous phyllite or schist fragment. Note the very coarse constituent-crystal size (far larger than normal clay size) and the strong crystal orientation. A number of similar, smaller MRF's also can be seen in this photo.

0.10 mm

Cretaceous Tuktu Fm.
Alaska

Same as previous view but with crossed polarizers. Shows the moderately high birefringence of the micas which make up this low-rank metamorphic rock fragment.

XN   0.10 mm

Carboniferous Sulzbacher Sch.
Germany

A rock composed almost entirely of metamorphic rock fragments. Elongate grain in center of field is a slate fragment or very low grade MRF. Other grains are from higher rank metamorphic rocks—schists, gneisses, and metaquartzites. Note orientation of component crystals parallel to long axis of slate fragment.

XN   0.27 mm

Cretaceous Fortress Mountain Fm.
Alaska

A high rank metamorphic rock fragment. Large grain in center shows composite, highly "stretched" quartz on the right and a vein of more equant, coarser quartz on the left. Although this grain is of metamorphic origin, it is surrounded by numerous detrital dolomite rhombs and rock fragments. This indicates a mixed metamorphic-sedimentary source area.

XN   0.15 mm

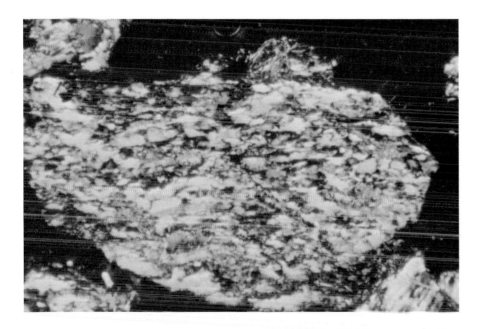

Oligocene Tongriano Fm.
Italy

A large grain of definite metamorphic origin. Consists of numerous, elongate, crenulate quartz crystals welded together. Most probably this is a fragment of a sheared metaquartzite.

XN   0.30 mm

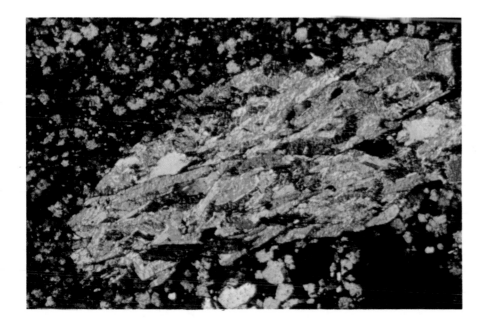

Triassic New Red Ss. ⋆
Northern Ireland

A detrital schist fragment (MRF). This grain consists almost entirely of muscovite crystals indicating that this metamorphic rock fragment was cut almost exactly parallel to schistosity along a micaceous layer.

XN   0.22 mm

Paleozoic andalusite schist
New Hampshire

This example shows a schistose texture commonly seen in high-rank metamorphic rock fragments. Elongate quartz grains are separated by thin mica plates. Detrital fragments of such rock types are normally quite soft and rarely survive extensive transport; however, when such fragments are found, they are excellent indicators of a metamorphic source.

XN   0.38 mm

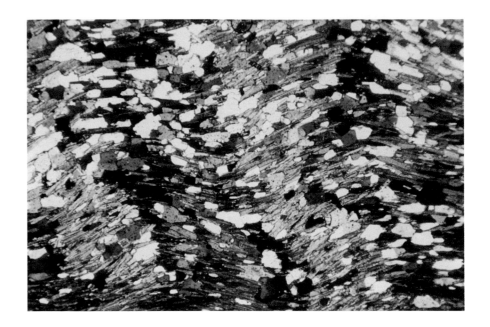

Upper Triassic New Haven Arkose
Connecticut

A detrital grain of composite or polycrystalline quartz of metamorphic origin. Because this grain consists entirely of quartz it is grouped with other quartz grains in many sandstone classifications. However, the nature of the quartz indicates a metamorphic origin and, in conjunction with other MRF's, it can be used to interpret source-area composition.

XN   0.15 mm

Cretaceous Torok Fm.
Alaska

An igneous rock fragment. This rounded, detrital grain contains quartz, feldspar, and other minerals. It is difficult to distinguish between an igneous or a metamorphic origin, but the coarse, equant crystals tend to favor an igneous (intrusive) origin.

XN   0.38 mm

Cretaceous Torok Fm.
Alaska

A probable igneous rock fragment. This grain could also have a high-rank metamorphic origin. Fragments of intrusive igneous rocks are generally difficult to distinguish from high-rank metamorphic rock fragments such as gneiss. Distinction must be made on the basis of recognizable textures if present; this is especially difficult in fine grained sediments.

0.38 mm

Cretaceous Torok Fm.
Alaska

Same as previous view but with crossed polarizers. Shows large quartz and feldspar crystals within the detrital rock fragment. The large variation in crystal sizes may favor an igneous rather than a metamorphic origin.

XN   0.38 mm

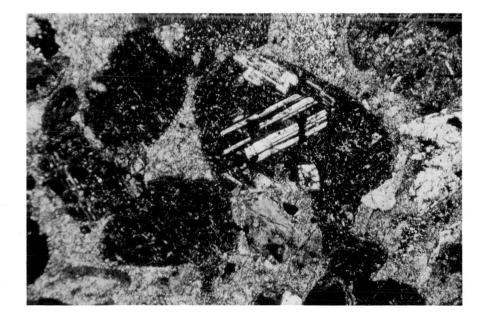

Cretaceous Ildefonso Fm. ★
Puerto Rico

Volcanic rock fragments (VRF's). The large grain in upper center shows laths of plagioclase set in a very finely crystalline matrix. The other dark grains are also VRF's but are much more difficult to identify because of the lack of phenocrysts. Such VRF's must be carefully distinguished from detrital chert or clay clasts. The cement in this example is calcite.

XN   0.30 mm

Cretaceous Shumagin Fm. *
Alaska

Volcanic rock fragments. Small laths of plagioclase feldspar and a very finely crystalline matrix make up most of these grains which are derived from basic, island-arc volcanic source areas. The small feldspar crystals are the main feature used to distinguish these grains from chert or clay fragments.

0.30 mm

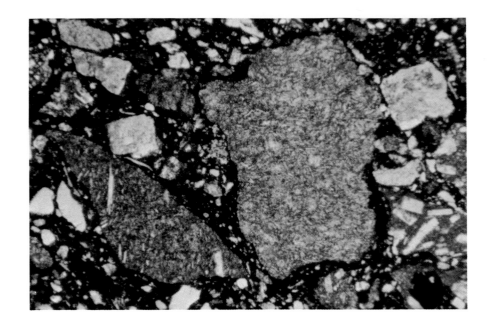

Cretaceous Shumagin Fm. *
Alaska

Same as previous view but with crossed polarizers. The VRF's consist of felted masses of very small, lath-like plagioclase crystals. Such VRF's are quite susceptible to destruction by abrasion or weathering but are excellent indicators of a volcanic-source terrain.

XN   0.30 mm

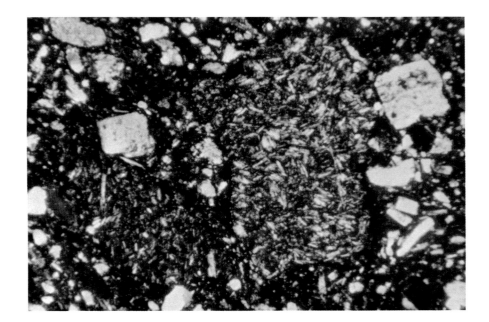

Tertiary Horse Springs Fm. *
Nevada

Abundant volcanic glass shards composed of opalline silica. Fragments of shard-filled sediment can be found and also are excellent indicators of a volcanic source area (primarily acidic volcanism).

0.10 mm

Tertiary Needles Range Fm.
Nevada

A volcanic feldspar grain. This grain would not be classified as a VRF, but it would be useful in confirming a volcanic source or in deciding whether associated fine-grained rock fragments are of volcanic origin. This example shows a plagioclase with well developed twinning, euhedral outline, and faint (but very diagnostic) compositional zoning set in a glassy groundmass.

XN   0.38 mm

Selected Rock Fragment Bibliography

Amsbury, D. L., 1962, Detrital dolomite in central Texas: Jour. Sed. Petrology, v. 32, p. 3-14.

Boggs, S., Jr., 1968, Experimental study of rock fragments: Jour. Sed. Petrology, v. 38, p. 1326-1339.

Cameron, K. L., and H. Blatt, 1969, Interpretive petrology of stream sediment, Elk Creek, Black Hills, South Dakota: Geol. Soc. America Abs. with Programs, v. 1, pt. 5, p. 12.

—— 1971, Durabilities of sand size schist and "volcanic" rock fragments during fluvial transport, Elk Creek, Black Hills, South Dakota: Jour. Sed. Petrology, v. 41, p. 565-576.

Dickinson, W. R., 1970, Interpreting detrital modes of graywacke and arkose: Jour. Sed. Petrology, v. 40, p. 695-707.

Sabins, F. F., Jr., 1962, Grains of detrital, secondary, and primary dolomite from Cretaceous strata of the Western Interior: Geol. Soc. America Bull., v. 73, p. 1183-1196.

Sneed, E. D., and R. L. Folk, 1958, Pebbles in the lower Colorado River, Texas; a study in particle morphogenesis: Jour. Geology, v. 66, p. 114-150.

Walton, A. W., 1977, Petrology of volcanic sedimentary rocks, Vieja Group, southern Rim Rock Country, Trans-Pecos Texas: Jour. Sed. Petrology, v. 47, p. 137-157.

Wolf, K. H., 1971, Textural and compositional transitional stages between various lithic grain types: Jour. Sed. Petrology, v. 41, p. 328-332.

Other Detrital Grains

Upper Cretaceous Navesink Fm. *
New Jersey

Glauconite, seen here as well rounded grains, is generally green in both ordinary and polarized light. Rounding does not necessarily indicate abrasion, and these grains were probably produced as fecal pellets *in situ*. Glauconite commonly occurs as pellets or within skeletal grains and, when not detrital, is a good indicator of a marine depositional setting.

XN   0.38 mm

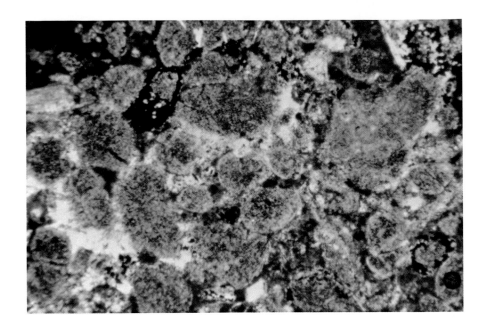

Upper Cretaceous Monte Antola Fm.
Italy

Detrital glauconite in a turbidite sandstone. These rounded glauconite grains were probably produced in shelf or slope sediments as pellets and were reworked into basinal turbidites. They show a darker, more typical, birefringence color than do the grains in the previous photo. Also note typical "granular" texture of glauconite.

XN   0.12 mm

Upper Cretaceous Tinton Sand
New Jersey

Large, unusual glauconite grains. The brown color is atypical and the crystal shape is clearly pseudomorphous after a precursor mineral such as vermiculite. Glauconite (celadonite) commonly forms as a degradation product of biotite or clay minerals in igneous and sedimentary rocks. These grains have undergone significant outcrop weathering.

0.38 mm

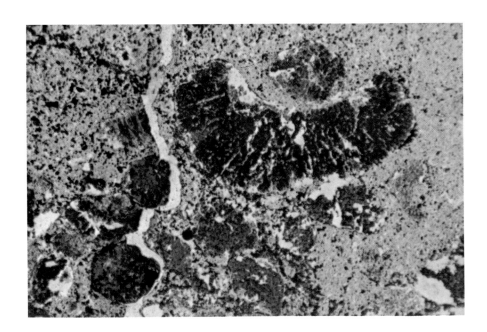

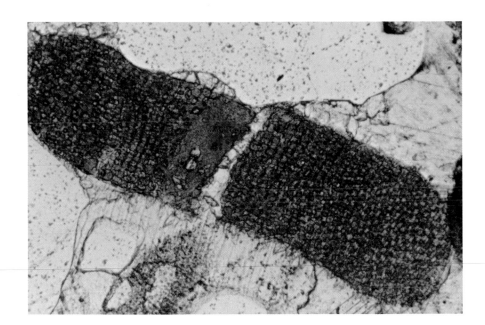

Jurassic Curtis Fm.
Utah

Authigenic glauconite found as an interstitial filling within the pores of an echinoderm fragment. The color is typical of glauconite; the texture, however, is typical of echinoid grains. In this example glauconite may have replaced some of the calcite of the echinoid.

0.27 mm

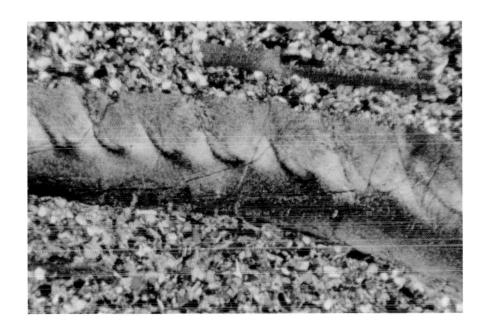

Oligocene Cyrenenschichten (Molasse)
Germany

A large calcium carbonate shell fragment in a siltstone. Shell can be identified by external morphology and internal shell structure as a pelecypod, *Cyrenia* sp. The presence of abundant shells of this mollusk but virtually no others aids in identification of the environment of deposition as being restricted marine.

XN   0.38 mm

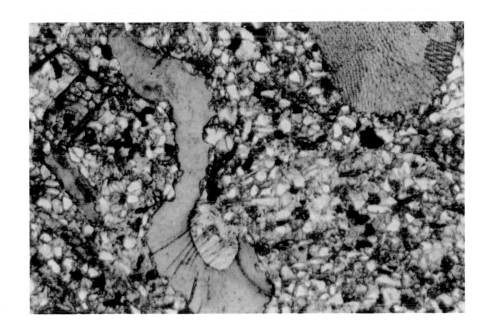

Ordovician Reedsville Shale
Pennsylvania

A varied assemblage of carbonate skeletal grains in a very fine grained sandstone. Visible are trilobite fragment (left center), bryozoan fragment (upper left), and crinoid plate (upper right). The composition and variety of the fauna aid in defining an open-marine depositional environment.

XN   0.38 mm

Upper Mississippian-Permian Nuka Fm.
Alaska

A carbonate shell fragment, probably molluscan, riddled with borings. The borings are especially visible in this grain because they are filled with hematite. Borings, especially algal borings, can commonly be used as indicators of shallow water (photic zone) origin of grains. The grains can, however, be transported subsequently into deeper water.

XN   0.24 mm

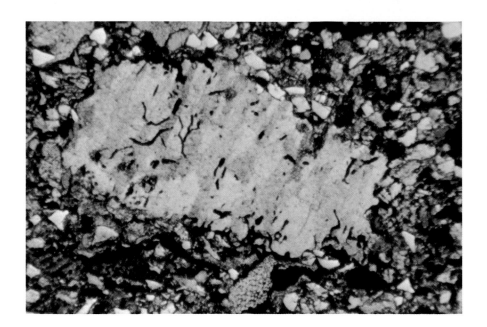

Lower Mississippian Cottonwood Canyon
    Mbr., Lodgepole Ls.
Montana

Phosphatic shell fragments, seen here, can also be significant constituents in some sediments. This example shows conodonts in a variety of sectioning angles. Note brownish tint to portions of these grains as well as complex external morphologies. Large brown grain at bottom is probably a phosphatic bone fragment.

                    0.38 mm

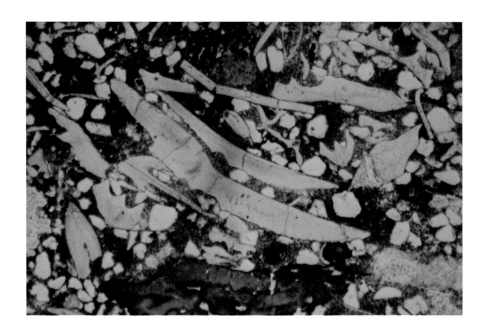

Lower Mississippian Cottonwood Canyon
    Mbr., Lodgepole Ls.
Montana

Same as previous view but with crossed polarizers. Shows typical extinction behavior of conodonts. The extremely low birefringence, and pronounced appearance of "white matter" within the sawtooth extinction pattern are characteristic.

XN   0.38 mm

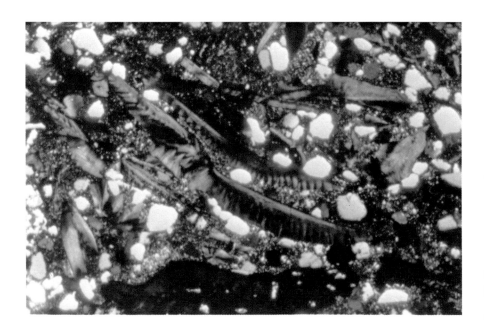

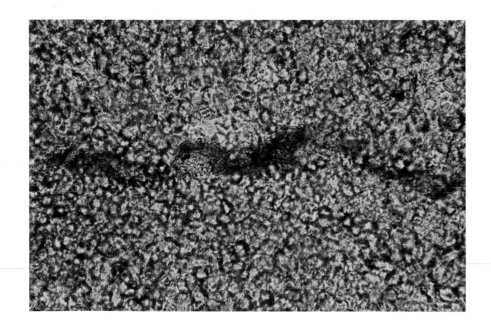

Tertiary Creede Fm.
Colorado

Organic matter. This is a deformed plant fragment. Most organic matter in older sediments consists of insoluble, largely indestructable compounds. The color of such material depends largely on its degree of thermal maturation. Matrix material in this sample is mainly calcite.

0.04 mm

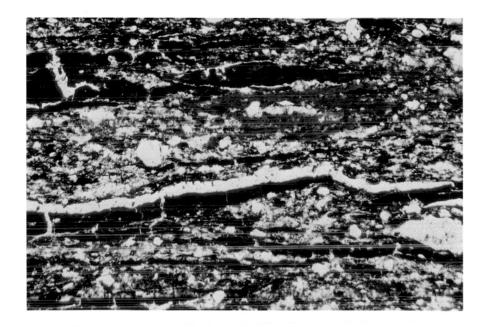

Upper Cretaceous Mesaverde Gp. ★
Colorado

Abundant organic matter in coaly seams. Note variation in color from deep reddish-brown to black—this is a function not only of thermal maturation but also of the type and thickness of organic detritus. The material seen here is derived from terrestrial, woody sources.

0.10 mm

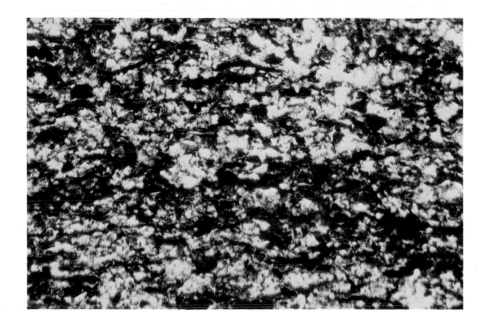

Upper Cretaceous Mesaverde Gp.
Colorado

Abundant disseminated blebs of organic matter throughout field of view. In this example the woody detritus has not formed discrete coaly seams but has remained as isolated grains or grain clusters partly deformed through compaction. Because of the opacity of this material it is normally best studied using reflected light or in specially prepared grain mounts.

0.10 mm

Pleistocene sandstone
California 646 m (2,120 ft)

Organic matter from the Salton Sea geothermal area is seen here in an unstained
grain mount. Note the range of colors
from medium brown to black largely depending on the thickness of the grains.
Pollen grains, such as the bisaccate form in
the lower right-center, generally are
medium to dark brown, indicative of
significant thermal maturation at the
150°C temperature encountered at this
depth.

0.10 mm

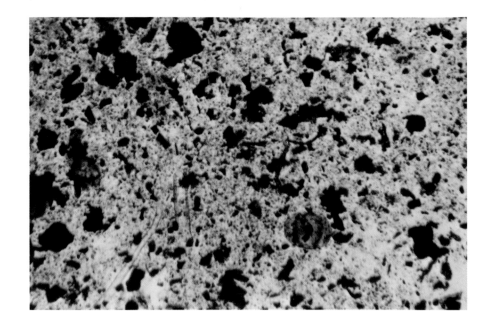

Pleistocene sandstone
California 1,033 m (3,390 ft)

Organic matter from the same Salton Sea
geothermal well as above example but
from a zone at 250°C (present-day temperature). Note that most grains, including
bisaccate pollen types, are virtually black,
indicating very advanced thermal maturation. This illustrates the utility of examination of organic matter in the determination
of former burial temperatures.

0.10 mm

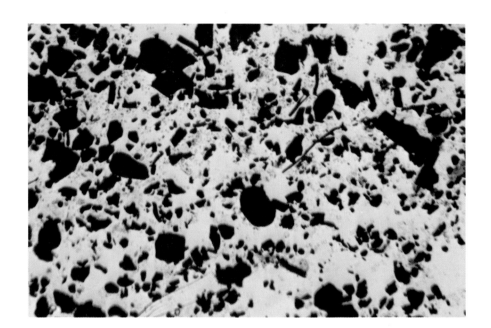

Quaternary gypsum dunes ★
Utah

Detrital discoidal gypsum crystals are a
common component of dune sands in
arid areas in which a source for gypsum is
present (the Bonneville Salt Flats in this
case). Such grains are only preserved where
rainfall is quite low or sedimentation
extremely rapid. Note partial etching of
grains and presence of carbonate inclusions
within gypsum, indicating that these
crystals grew within carbonate sediment.

XN   0.22 mm

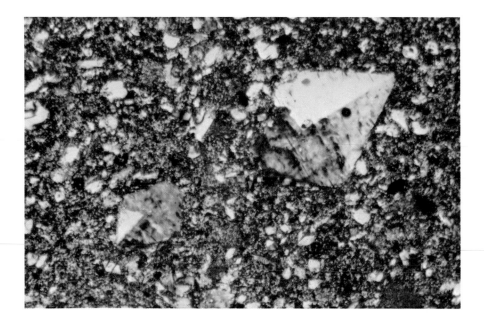

Miocene (Messinian) Gesso Solfifera
Italy

Detrital twinned gypsum crystals in a marine sediments. These crystals were preserved in a mass flow deposited in relatively deep waters. Note that gypsum crystals are floating in a matrix of fine-grained calcite and show little indication of corrosion.

XN   0.17 mm

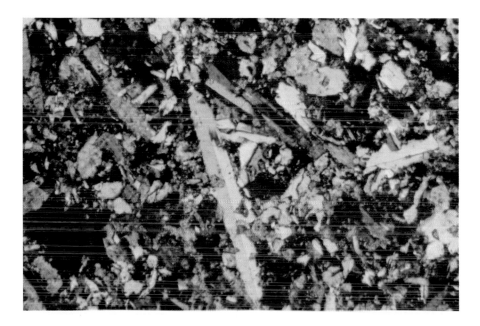

Miocene (Messinian) Gesso Solfifera
Italy

Detrital needle- or lath-shaped crystals of gypsum preserved in a marine turbidite. The crystals, which are twinned, were presumably deposited in shallow water and were reworked into a deeper water setting by turbidity currents. Numerous crystal fragments of gypsum and calcite make up the rock matrix.

XN   0.22 mm

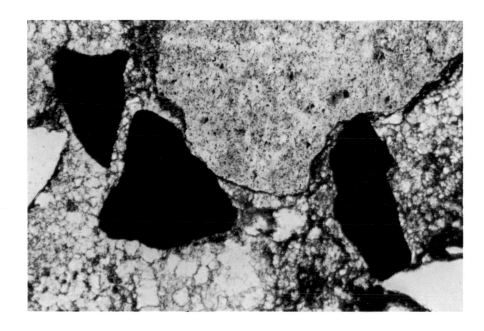

Permian Abo Ss.
New Mexico

Detrital heavy minerals can be abundant in sedimentary rocks, especially where concentrated in placers. The iron oxides seen here (probably magnetite with some hematitic alteration) are opaque in transmitted light and are best examined in reflected light. Many opaque grains are of authigenic origin, however, and thin section study is very useful in distinguishing authigenic from detrital grains.

0.27 mm

Holocene sediment ★
Connecticut

Detrital magnetite grains; a reflected-light photo of a grain mount. Identification of heavy minerals is facilitated, in many cases, by crushing the rock and concentrating the "heavies" by heavy liquids or by other types of mineral-separation techniques. Separated grains can be examined in refractive index oils or in polished mounts. Magnetite shows metallic, silver-black color when seen in reflected light.

RL   0.06 mm

Upper Triassic Brunswick Fm.
New Jersey

A pyrite crystal in reflected light. Although this grain is authigenic, rather than detrital, it shows the typical yellow-gold metallic color of pyrite. Speckled texture is due to plucking of material during section grinding and is minimized in polished sections.

RL   0.22 mm

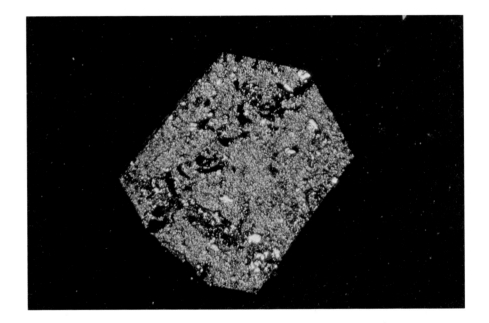

Jurassic Eisenoolith
Germany

Hematitic ooids illuminated with very strong transmitted light (conoscopic condenser lens inserted). Note the reddish-yellow color indicative of hematite which may be partially altered to goethite-limonite. Oolitic iron minerals include hematite, chamosite, limonite, and siderite.

0.24 mm

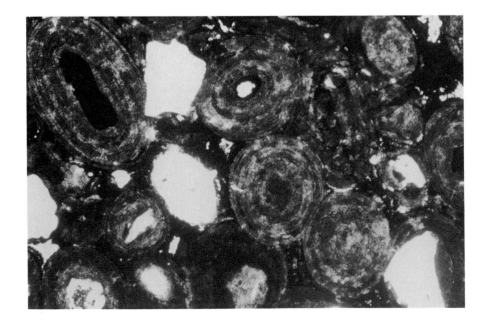

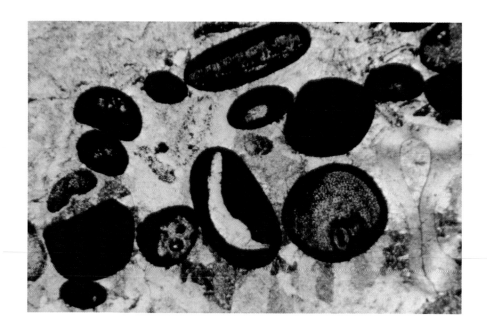

Silurian Clinton Fm.
Pennsylvania

Hematite, here as oolitic coatings on carbonate skeletal fragments. The hematite in this sample is opaque, but in very thin sections, with stronger transmitted light, or in reflected light, one can commonly distinguish a dark red to brown color characteristic of this mineral. Hematitic ooids are normally indicative of oxidizing marine environments, as well as paleosoils and weathering horizons.

XN   0.38 mm

Cretaceous Monte Antola Fm. *
Italy

Detrital micas (muscovite). The grains with the bright blue (second order) birefringence are muscovite flakes. They are nearly colorless in ordinary light. The slightly speckled texture (reminiscent of birch bark) is characteristic of micas. Muscovite, because of its greater chemical stability, is more common than biotite in most sedimentary rocks.

XN   0.10 mm

Ordovician(?) garnet schist
Connecticut

A large biotite crystal surrounded by muscovite. Note brown color, excellent cleavage, and dark spots which are "pleochroic halos" formed around minute inclusions of zircon, apatite, or other uranium-bearing minerals. Biotite crystals are normally pleochroic, with colors ranging from colorless to yellow, brown, red-brown, and green. Biotite weathers readily and if very abundant in a sediment one can suspect a volcanic source.

0.30 mm

Cambrian(?) Hitchcock Lake Mbr. of
Waterbury Gneiss *
Connecticut

A biotite grain surrounded by quartz. The
biotite shows third order birefringence
colors and the rough (birch bark) texture
typical of micas. Two small pleochroic
halos are also visible. Biotite can be derived
in small amounts from almost all types of
igneous and metamorphic terrains.

XN   0.30 mm

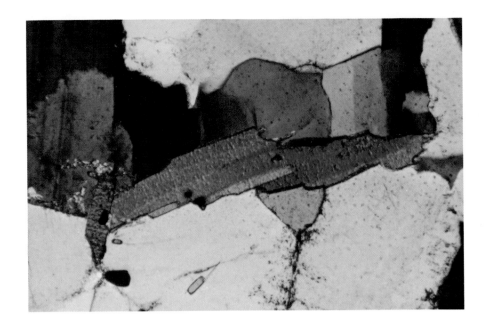

Ordovician(?) garnet schist
Connecticut

Biotite being altered to chlorite during
retrograde metamorphism. The alteration
of biotite to chlorite can also take place as
a weathering reaction and all stages of
transition can be found. The green-colored
grains are now largely chlorite; the brown
ones are biotite. Note relict pleochroic
halos within the chloritized biotites.

0.24 mm

Ordovician(?) garnet schist
Connecticut

Same as previous view but with crossed
polarizers. Note "ultra blue" anomalous
birefringence colors of the chlorite and the
damaged zones comprising the pleochroic
halos (remnants from the pre-existing biotite). The unaltered biotite flakes show
high birefringence. Chlorite can form both
as an alteration of biotite or as a "primary"
mineral in low grade schists and phyllites.

XN   0.24 mm

Permian Abo Ss.
New Mexico

A detrital chlorite grain. Shows anomalous birefringence colors (some chlorite varieties have "ultra blue" colors; others have more normal low birefringence). The coarseness of this chlorite grain indicates that it probably is an alteration of biotite. Chlorite can be distinguished from clinozoisite (which also has "ultra blue" birefringence) by the higher relief of the latter mineral. Chlorite is found in most source rock types.

XN   0.06 mm

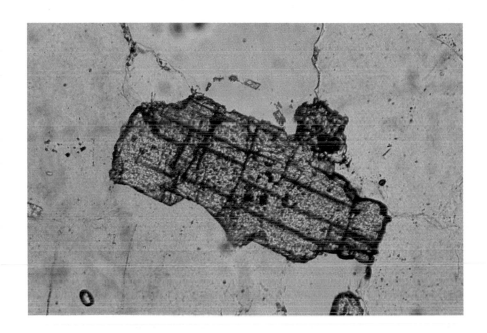

Ordovician(?) schist
Connecticut

A kyanite crystal showing typical high relief, long and bladed crystal form, two good cleavages, and light color (often is pleochroic). Kyanite is only found in high-grade metamorphic source areas and thus is a valuable provenance indicator. It has moderate chemical stability but relatively low abrasion resistance. Many varieties commonly make it useful for stratigraphic correlation.

0.10 mm

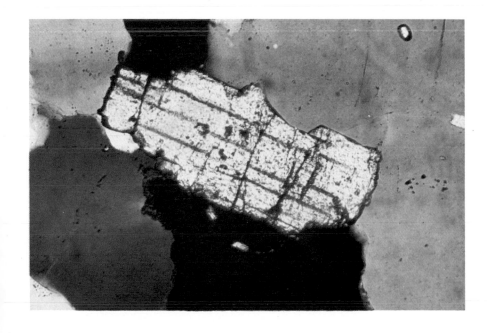

Ordovician(?) schist
Connecticut

Same as previous view but with crossed polarizers. Kyanite is shown surrounded by quartz. The birefringence colors seen here are well below the maximum shown by kyanite (first-order orange to red). Characteristic features include high relief, moderate to low birefringence, pleochroism, and a biaxial negative figure with a 2V of 82°.

XN   0.10 mm

Ordovician-Silurian schist
Massachusetts

Sillimanite. The fibrous crystal form, pale brown color, slight pleochroism, and high relief are characteristic. Sillimanite is found only in metamorphic rocks (mainly high-grade schists and contact metamorphics). This mineral has moderate chemical stability and relatively low abrasion resistance.

0.10 mm

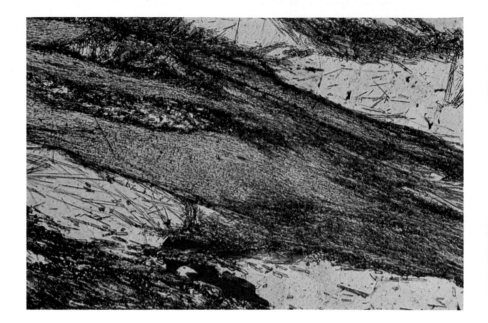

Ordovician-Silurian schist
Massachusetts

Same as previous view but with crossed polarizers. Under polarized light sillimanite is characterized by moderate birefringence, parallel extinction, positive elongation, a biaxial positive figure, and a 2V of 24°. Larger crystals are common; this needle-like variety is termed fibrolite.

XN   0.10 mm

Paleozoic andalusite schist
New Hampshire

A large andalusite crystal with excellent 110 cleavage surrounded by muscovite. Andalusite is characterized by high relief, color ranging from colorless to pink (occasionally green, or yellow), variable pleochroism, and excellent cleavage. It is most common in schists and contact metamorphic rocks. Low chemical stability in surface environments explains its scarcity in older sediments; rather common in younger units, however.

0.30 mm

Paleozoic andalusite schist
New Hampshire

Same as previous photo but with crossed polarizers. Andalusite has low birefringence, negative elongation, a biaxial negative figure, and a 2V of 80°-85°. The mineral is also characterized by abundant inclusions, as with the muscovite crystals (high birefringence) seen here.

XN   0.30 mm

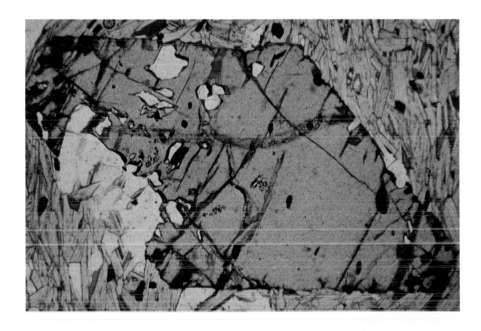

Ordovician schist
Connecticut

A staurolite crystal surrounded by quartz (colorless) and muscovite (stained red in this section). Staurolite has brownish color, moderate relief, moderate pleochroism, abundant inclusions, and prismatic crystal habit with weakly developed cleavage. It is an excellent indicator of a schistose metamorphic source. Detrital grains are rarely well crystallized.

0.38 mm

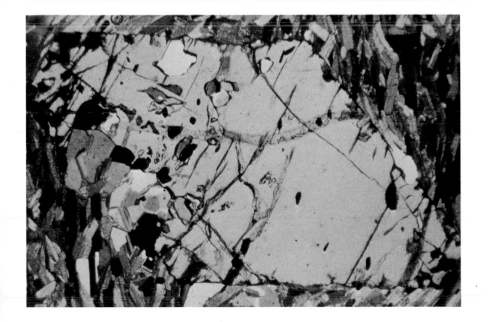

Ordovician schist
Connecticut

Same as previous view but with crossed polarizers. Staurolite normally shows moderate to strong birefringence, positive elongation, a biaxial positive figure, and a 2V of 80°-90°. Detrital specimens commonly show some surficial alteration to chlorite.

XN   0.38 mm

Ordovician Straits Schist
Connecticut

Hornblende, shown here, is a monoclinic amphibole with brownish to greenish color, commonly with strong pleochroism. Relief is high and cleavages are very well developed and intersect at 124°. Hornblende is derived from igneous and metamorphic rocks, especially granite, syenite, diorite, and amphibolite. Surficial alteration to chlorite is common.

0.38 mm

Ordovician Straits Schist
Connecticut

Same as previous view but with crossed polarizers. Hornblende is characterized by moderate birefringence (second-order colors), positive elongation, a biaxial negative figure, and a 2V of 70°-85°. The example shown here has less than maximum birefringence colors.

XN   0.38 mm

Lower Miocene Arikaree Fm.
Wyoming

Detrital green hornblende crystals derived from a metamorphic-granitic source. Green hornblendes are by far the most common as detrital grains. They tend to have variable pleochroism with some grains showing strong effects, most showing moderate pleochroism, and a few showing virtually none. Detrital grains are often found as elongate cleavage fragments with frayed ends due to dissolution.

0.08 mm

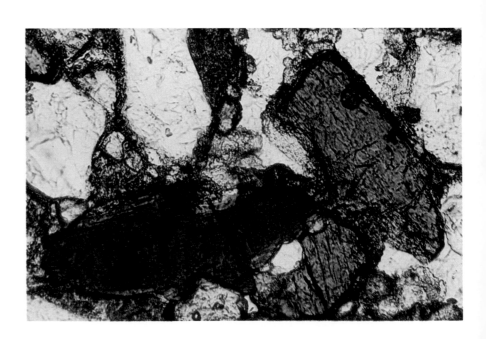

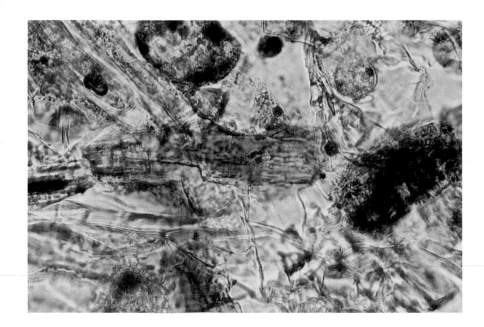

Lower Miocene Arikaree Fm.
Nebraska

A detrital hornblende grain with light-brown color in this orientation. Grain is derived from a volcanic source and is surrounded by volcanic glass shards. Compare with next photograph.

0.27 mm

Lower Miocene Arikaree Fm.
Nebraska

Approximately the same field of view as previous photo but rotated 90°. Note the extreme color change of the detrital hornblende compared with previous photo. Brown hornblende crystals are relatively uncommon in sediments. Such extreme pleochroism is generally found in Na-rich hornblende.

0.27 mm

Devonian(?) meta-diabase
Connecticut

A pyroxene crystal, probably pigeonite, surrounded by plagioclase feldspars. This is one of many pyroxene types (augite, diopside, enstatite, hypersthene, and spodumene) which may be found as detrital grains in sedimentary rocks. All have high relief, characteristic strong cleavages which intersect at 87°-93°, short prismatic crystals, light color, and weak pleochroism. Pyroxenes are derived mainly from igneous and some metamorphic rocks.

0.10 mm

Devonian(?) meta-diabase
Connecticut

Same as previous view. In polarized light pigeonite shows moderate to strong birefringence (as seen in lower right), a biaxial positive figure, and a 0°-20° 2V. Other pyroxenes, such as enstatite and hypersthene have much lower birefringence colors; most are optically positive. Pigeonite has parallel extinction and is length positive. Generally uncommon in sediments due to chemical instability.

XN   0.10 mm

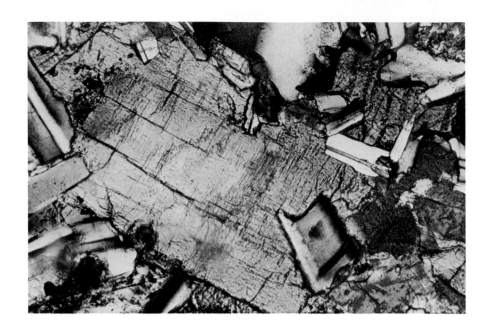

Cambrian Potsdam Ss.
New York

Tourmaline crystals in a limestone (tourmaline formed as a hydrothermal alteration product). Under polarized light, tourmaline shows moderate to strong birefringence, a uniaxial figure (often indeterminate; when strained, may appear biaxial), and negative elongation. This is one of the commonest heavy minerals in sediments because of its chemical and physical stability. Rarely shows alteration.

XN   0.24 mm

Cambrian Gatesburg Fm.
Pennsylvania

A detrital tourmaline with an authigenic overgrowth. Shows very poor cleavage, high relief, green to brown color with extreme pleochroism (goes completely black with rotation of stage), and oriented, ragged overgrowths. Tourmaline is derived from acidic igneous rocks and many types of metamorphic rocks, both regional and contact.

0.025 mm

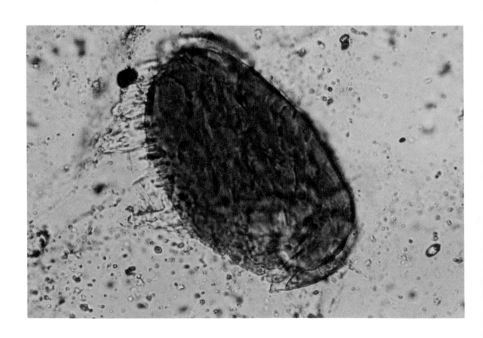

Upper Triassic arkose
Massachusetts

A detrital tourmaline grain, probably of first-cycle origin from a metamorphic-granitic terrain. The range of tourmaline color and pleochroism types makes it possible to use this mineral as a marker in correlation studies.

XN   0.08 mm

Precambrian Ortega Quartzite
New Mexico

Epidote crystals in a quartz vein in a metamorphic terrain. Epidote shows high relief, gray-brown color (also commonly green), perfect basal cleavage, and weak pleochroism. Epidote is derived mainly from altered igneous rocks and crystalline metamorphic terrains.

0.27 mm

Precambrian Ortega Quartzite
New Mexico

Same as previous photo but with crossed polarizers. Epidote has moderate to strong birefringence, a biaxial negative figure, and a 2V of 92°. Detrital grains are commonly irregular and angular with very uneven fracture patterns.

XN   0.27 mm

54

Precambrian Inwood Ls.
New York

Sphene crystals in a calcitic marble. Sphene
(sometimes called titanite) shows characteristic diamond- or wedge-shaped crystal
outlines with good cleavages, and a brownish-yellow color. Relief is very high; pleochroism is very slight except in strongly
colored varieties. Sources include granites,
intermediate igneous rocks, and metamorphics (gneisses, schists, and altered
limestones). Can also be authigenic.

0.24 mm

Precambrian Inwood Ls.
New York

Same as previous photo but with crossed
polarizers. Under polarized light, sphene
has extreme birefringence (compare with
adjacent calcite which also has extreme
birefringence). Sphene also has a biaxial
positive figure with a 2V of 23°-34° and
very strongly inclined dispersion.

XN   0.24 mm

Oligocene Fish Canyon Tuff
Colorado

Apatite crystals as inclusions in biotite.
Hexagonal outline, imperfect cleavage, high
relief, very low birefringence, uniaxial
negative figure, and weak to absent pleochroism characterize apatite. A moderately
stable mineral, it is commonly found as
detrital grains in sedimentary rocks.
Derived mainly from igneous rocks,
especially granites and syenites.

XN   0.06 mm

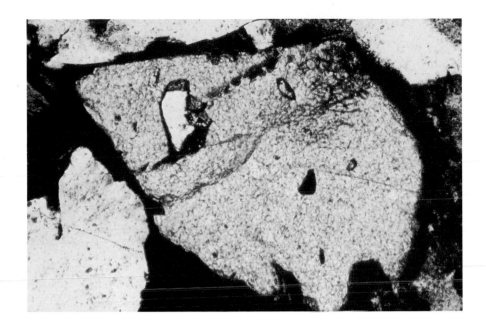

Upper Triassic New Haven Arkose
Connecticut

A detrital garnet, possibly almandite. Garnet is among the easiest minerals to identify because of its very high relief coupled with poor cleavage, a characteristic pitted surface texture in thin section and, most importantly, its isotropism under cross-polarized light. This example is surrounded by hematite cement. Garnets may be derived from igneous and/or metamorphic sources, and almandite is the most common detrital garnet.

0.08 mm

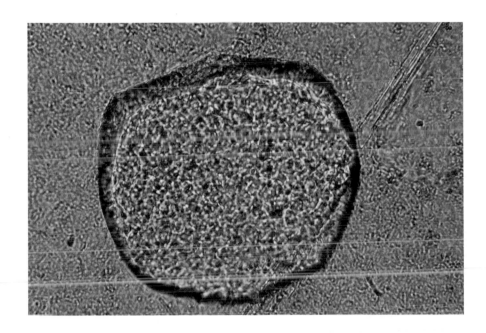

Cambrian(?) Hitchcock Lake Mbr. of
    Waterbury Gneiss
Connecticut

A single garnet crystal, probably grossularite, showing a primary rounded outline, very high relief, and a pitted surface texture produced during grinding of the thin section. The color of garnets can range from colorless to red, with scarcer varieties being green or yellow. Garnets, although not particularly stable, are often abundant in sedimentary rocks.

0.04 mm

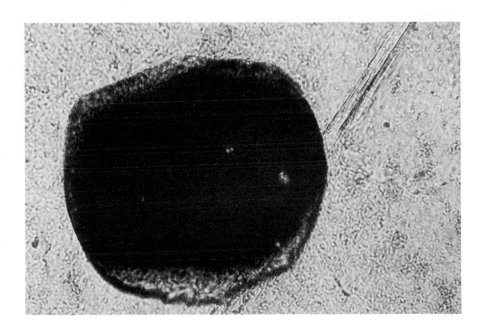

Cambrian(?) Hitchcock Lake Mbr. of
    Waterbury Gneiss
Connecticut

Same as previous photo but with crossed polarizers. Shows complete isotropism of garnet (slight birefringent effects along grain margins are probably due to overlap of adjacent birefringent quartz crystals). The only common detrital mineral which has optical properties similar to garnet (especially to almandite) is spinel.

XN   0.04 mm

Paleozoic schist
Massachusetts

Probable clinozoisite crystals. Clinozoisite (a member of the epidote group) shows pale green color in this example but is commonly colorless; it also has high relief, pronounced cleavage, and inclusions which parallel the c-axis. Clinozoisite is relatively common in sediments although it is frequently mistaken for zoisite. Derived mainly from schists and metamorphosed basic volcanic rocks.

0.08 mm

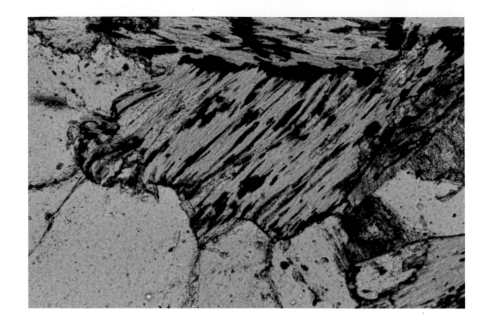

Paleozoic schist
Massachusetts

Same as previous photo but with crossed polarizers. Shows anomalous "ultra blue" birefringence (not present in all clinozoisite grains). It has a biaxial positive figure with a 2V near 90°, positive or negative elongation, and low birefringence. Clinozoisite has slightly oblique extinction which distinguishes it from zoisite.

XN   0.08 mm

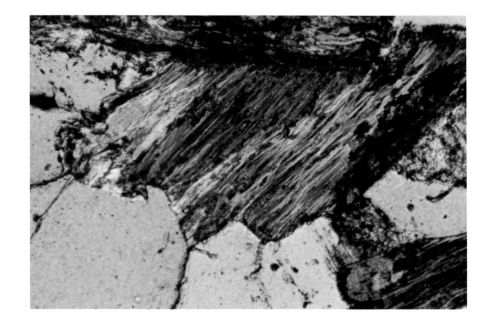

Jurassic Morrison Fm.
Utah

A rounded, detrital zircon crystal. Zircon commonly shows an abraded bipyramidal prismatic form, generally is colorless, and has extremely high relief. Cleavage is imperfect to poor, and some crystals show slight pleochroism. Most zircons are colorless; some are pale yellow, brown, pink, or purple. Zircon is a ubiquitous component of sedimentary rocks because of its great physical and chemical stability.

0.03 mm

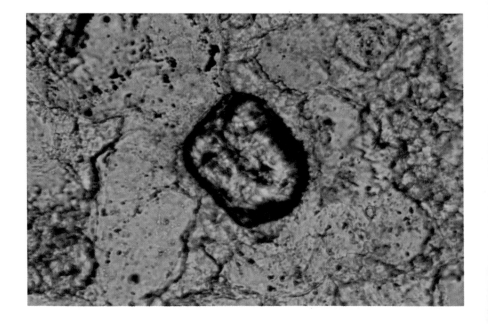

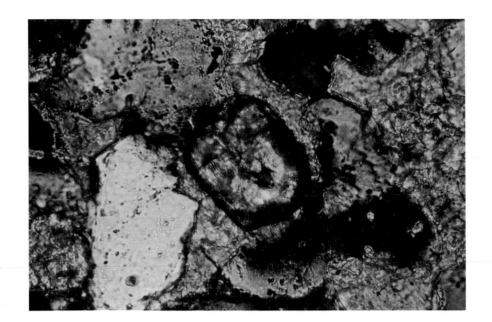

Jurassic Morrison Fm. ★
Utah

Same as previous view. Under cross-polarized light zircon shows strong to extreme birefringence, a uniaxial positive figure, and positive elongation. Virtually never shows alteration but inclusions are common. Mainly derived from acidic to intermediate igneous rocks (less common from metamorphics). Can be confused with xenotime.

XN   0.03 mm

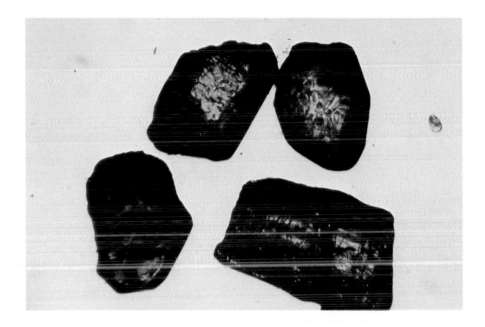

Eocene Green River Fm.
Colorado

Although heavy minerals can be readily identified in thin section, a better idea of the total suite of heavy minerals in a sample can be obtained by using mineral separations. Individual grains can still be examined with the microscope, can be placed in refractive index oils, or studied by X-ray. This example shows epidote crystals. Color, crystal outline, surface features, and weathering-abrasion effects are important criteria for grain identification.

0.17 mm

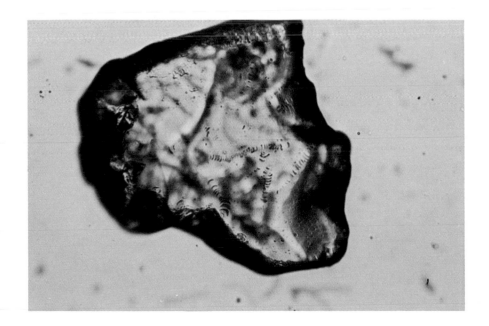

Holocene sediment
Minnesota

A mineral separate of garnet. Note lack of cleavage, irregular to conchoidal fracture, and pale pink-yellow color which are characteristic of garnet. Also visible are trails of chatter marks which have been interpreted as being indicative of glacial transport (Folk, 1975).

0.05 mm

Holocene sediment
Minnesota

A mineral separate of garnet. This grain, in addition to showing the typical characteristics of garnet, also shows extensive chemical etching which has yielded the pitted, V-shaped surface features. Such dissolution can take place either during transportation and deposition of grains, or intrastratally during diagenesis. In many cases, remarkably fragile, deeply etched grains can be found in sediments.

0.05 mm

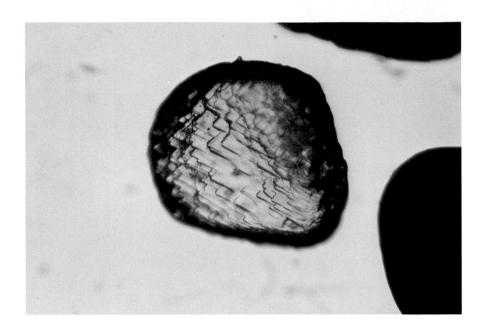

Holocene beach sediment
Connecticut

A mineral separate of hornblende derived from Paleozoic metamorphics. Note the irregular grain shape only partly controlled by cleavage, as well as the variation in color intensity from the grain margin to the center—a function of the variation in grain thickness from edge to center. Grain is identifiable by color, pleochroism, shape, and birefringence.

0.08 mm

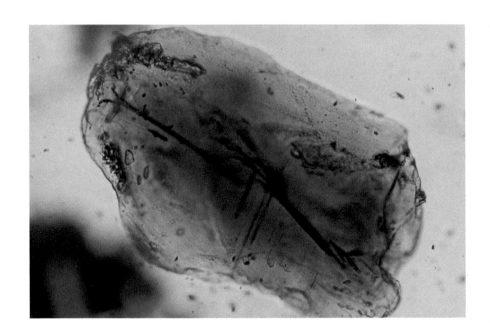

Holocene beach sediment
Connecticut

Same as previous photo but with crossed polarizers. Birefringence colors are much higher than for hornblende in thin section. This is due to the much greater thickness of the grain in this mount than the standard 30 μm of a thin section. Variation of birefringence (color banding) around margins reflects changes of grain thickness from the edge to the center of the grain.

XN   0.08 mm

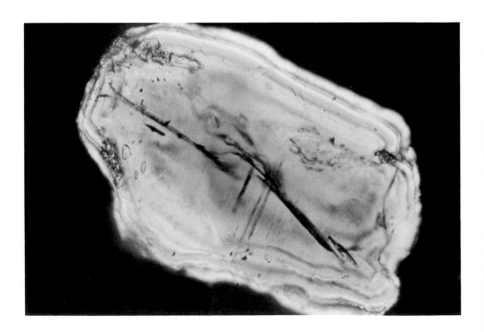

Holocene beach sediment ★
Connecticut

A kyanite crystal in grain mount. The high relief, excellent cleavage, bladed habit, and color (including pleochroism) allow identification of this mineral as kyanite. The angularity and lack of alteration indicate that the grain is first-cycle and has probably undergone little chemical weathering and short transport.

0.08 mm

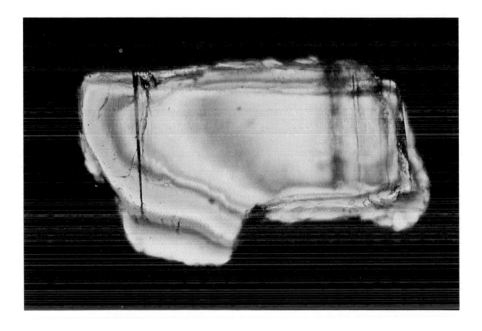

Holocene beach sediment ★
Connecticut

Same as previous photo but with crossed polarizers. Moderate birefringence aids in identification of this mineral as kyanite. Birefringence, however, is difficult to use, as exact thickness of grain is not known. In fact, the thickness of the left side of the grain is considerably less than that of the right side as can be seen from the color-banding patterns.

XN   0.08 mm

Eocene Green River Fm.
Colorado

Rutile crystals in grain mount. Yellow to red-brown color, striations, twinning (seen in grain at upper left), moderate pleochroism, and prismatic outline with pyramidal terminations are typical for rutile. This is an extremely common detrital heavy mineral derived mainly from acidic igneous rocks, crystalline metamorphic rocks, or in situ decomposition of ilmenite.

0.10 mm

60

Eocene Green River Fm.
Colorado

Sphene (titanite) crystals in grain mount.
Note brownish-yellow color, abundant
inclusions, and abrasional rounding.

0.17 mm

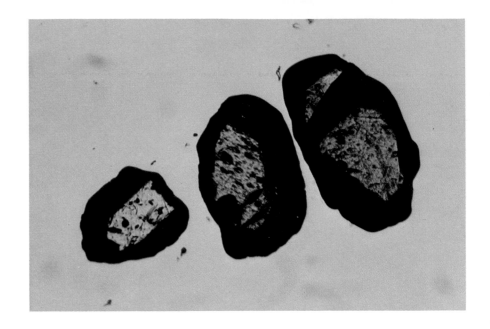

Eocene Green River Fm.
Colorado

Tourmaline crystals in grain mount.
Prismatic crystals show pronounced external striations, strong pleochroism, and
red-brown to green color.

0.17 mm

Eocene Green River Fm. ★
Colorado

Zircon crystals in grain mount. Euhedral
crystals with bipyramidal terminations,
high relief, very pale-yellow to pink color,
and common inclusions are characteristic
of zircon.

0.22 mm

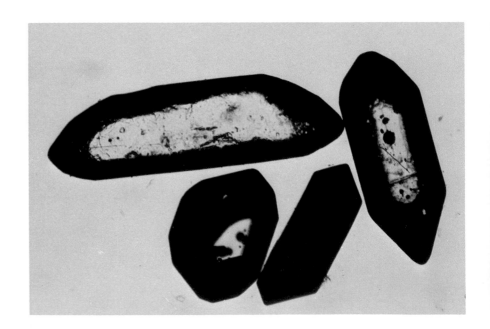

Selected Detrital Grain Bibliography

Glauconite

Boyer, P. S., E. A. Guiness, M. A. Lynch-Blosse, and R. A. Stolzman, 1977, Greensand fecal pellets from New Jersey: Jour. Sed. Petrology, v. 47, p. 267-280.

Burst, J. F., 1958, "Glauconite" pellets: their mineral nature and application to stratigraphic interpretation: AAPG Bull., v. 42, p. 310-327.

Cimbálniková, A., 1971, Chemical variability and structural heterogeneity of glauconites: Am. Mineralogist, v. 56, p. 1385-1398.

Cloud, P. E., 1955, Physical limits of glauconite formation: AAPG Bull., v. 39, p. 484-492.

Ehlmann, A. J., N. C. Hulings, and E. D. Glover, 1963, Stages of glauconite formation in modern foraminiferal sediments: Jour. Sed. Petrology, v. 33, p. 87-96.

Gruner, J. W., 1935, The structural relationship of glauconite and mica: Am. Mineralogist, v. 20, p. 699-714.

Hein, J. R., A. O. Allwardt, and G. B. Griggs, 1974, The occurrence of glauconite in Monterey Bay, California: diversity, origins, and sedimentary environmental significance: Jour. Sed. Petrology, v. 44, p. 562-571.

Porrenga, D. H., 1967, Glauconite and chamosite as depth indicators in the marine environment: Marine Geology, v. 5, p. 495-501.

Triplehorn, D. M., 1966, Morphology, internal structure, and origin of glauconite pellets: Sedimentology, v. 6, p. 247-266.

Carbonates

Bathurst, R. G. C., 1971, Carbonate sediments and their diagenesis: New York, Elsevier, 620 p.

Gubler, Yvonne, J. P. Bertrand, L. Mattavelli, A. Rizzini, and R. Passega, 1967, Petrology and petrography of carbonate rocks, in G. V. Chilingar, H. J. Bissell, and R. W. Fairbridge, eds., Carbonate rocks, Devel. in Sed. 9A: New York, Elsevier, v. A, p. 51-86.

Johnson, J. H., 1951, An introduction to the study of organic limestones: Colo. School of Mines Quart., v. 46, no. 2, 185 p.

Lippmann, F., 1973, Sedimentary carbonate minerals: New York, Springer-Verlag, 228 p.

Majewske, O. P., 1969, Recognition of invertebrate fossil fragments in rocks and thin sections: Leiden, E. J. Brill, 101 p., 106 pl.

Milliman, J. D., 1974, Marine carbonates: New York, Springer-Verlag, 375 p.

Pilkey, O. H., B. W. Blackwelder, L. J. Doyle, and E. L. Estes, 1969, Environmental significance of the physical attributes of calcareous sedimentary particles: Gulf Coast Assoc. Geol. Socs., Trans., v. 19, p. 113-114.

Sabins, F. F., Jr., 1962, Grains of detrital, secondary, and primary dolomite from Cretaceous strata of the Western Interior: Geol. Soc. America Bull., v. 73, p. 1183-1196.

Scholle, P. A., 1978, A color illustrated guide to carbonate rock constituents, textures, cements, and porosities: AAPG Memoir 27, 241 p.

Swinchatt, J. P., 1969, Algal boring: a possible depth indicator in carbonate rocks and sediments: Geol. Soc. America Bull., v. 80, p. 1391-1396.

Wilson, J. L., 1975, Carbonate facies in geologic history: New York, Springer-Verlag, 471 p.

Iron Minerals

Edwards, A. B., 1958, Oolitic iron formations in northern Australia: Geol. Rundschau, v. 47, p. 668-682.

Gottesmann, B., 1966, Mineralogisch-petrographische Untersuchungen an den oolithischen Eisenerzen des Jura von Sommerschenburg westlich Magdeburg: Geologie, v. 15, p. 309-339.

Knox, R. W. O'B., 1970, Chamosite ooliths from the Winter Gill ironstone (Jurassic) of Yorkshire, England: Jour. Sed. Petrology, v. 40, p. 1216-1225.

Organic Matter

Ammosov, I. I., 1968, Coal organic matter as a parameter of the degree of sedimentary rock lithification: 23rd Internat. Geol. Cong. Proc., Prague, v. 11, p. 23-30.

Bostick, N. H., 1971, Thermal alteration of clastic organic particles as an indicator of contact and burial metamorphism in sedimentary rocks: Geoscience and Man, v. 3, p. 83-92.

—— and B. Alpern, 1977, Principles of sampling, preparation and constituent selection for microphotometry in measurement of maturation of sedimentary organic matter: Jour. Microscopy, v. 109, pt. 1, p. 41-47.

Brooks, J. D., 1970, The use of coals as indicators of the occurrence of oil and gas: Australian Petroleum Exploration Assoc. Jour., v. 10, pt. 2, p. 35-40.

Hood, A., and J. R. Castaño, 1974, Organic metamorphism: its relationship to petroleum generation and application to studies of authigenic minerals: U.S. Econ. Comm. Asia Far East, Comm. Coord. Joint Prospect. Miner. Resour. Asian Offshore Areas. Tech. Bull., v. 8, p. 85-118.

—— C. C. M. Gutjahr, and R. L. Heacock, 1975, Organic metamorphism and the generation of petroleum: AAPG Bull., v. 59, p. 986-996.

Smith, M. H., 1970, Identification of organic matter in thin section by staining and a study programme for carbonate rocks: Jour. Sed. Petrology, v. 40, p. 1350-1351.

Staplin, F. L., 1969, Sedimentary organic matter, organic metamorphism, and oil and gas occurrence: Canad. Petrol. Geol. Bull., v. 17, p. 47-66.

Heavy Minerals

Andel, T. H. van, 1959, Reflections on the interpretation of heavy mineral analyses: Jour. Sed. Petrology, v. 29, p. 153-163.

Blatt, Harvey, and Berry Sutherland, 1969, Intrastratal solution and non-opaque heavy minerals in shales: Jour. Sed. Petrology, v. 39, p. 591-600.

Bornhauser, M., 1940, Heavy mineral associations in Quaternary and late Tertiary sediments of the Gulf Coast of Louisiana and Texas: Jour. Sed. Petrology, v. 10, p. 125-135.

Boswell, P. G. H., 1933, On the mineralogy of the sedimentary rocks: London, Thomas Murby and Co., 393 p.

Bradley, J. S., 1957, Differentiation of marine and sub-aerial sedimentary environments by volume percentage of heavy minerals; Mustang Island, Texas: Jour. Sed. Petrology, v. 27, p. 116-125.

Caroll, D., R. B. Neuman, and H. W. Jaffe, 1957, Heavy minerals in arenaceous beds in parts of the Ocoee Series, Great Smoky Mountains, Tennessee: Am. Jour. Sci., v. 255, p. 175-193.

Cogen, W. M., 1940, Heavy-mineral zones of Louisiana and Texas Gulf Coast sediments: AAPG Bull., v. 24, p. 2069-2101.

Feo-Codecido, G., 1956, Heavy-mineral techniques and their application to Venezuelan stratigraphy: AAPG Bull., v. 40, p. 948-1000.

Folk, R. L., 1975, Glacial deposits identified by chattermark trails in detrital garnets: Geology, v. 3, p. 473-475.

Heimlich, R. A., L. B. Shotwell, T. Cookro, and M. J. Gawell, 1975, Variability of zircons from the Sharon Conglomerate of northeastern Ohio: Jour. Sed. Petrology, v. 45, p. 629-635.

Hsü, K. J., 1960, Texture and mineralogy of the Recent sands of the Gulf Coast: Jour. Sed. Petrology, v. 30, p. 380-403.

Hutton, C. O., 1950, Studies of heavy detrital minerals: Geol. Soc. America Bull., v. 61, p. 635-716.

Krynine, P. D., 1946, The tourmaline group in sediments: Jour. Geology, v. 54, p. 65-87.

Ledent, D., C. Patterson, and G. R. Tilton, 1964, Ages of zircon and feldspar concentrates from North American beach and river sands: Jour. Geology, v. 72, p. 112-122.

Mackie, William, 1923, The principles that regulate the distribution of particles of heavy minerals in sedimentary rocks, as illustrated by the sandstones of the northeast of Scotland: Edinburgh Geol. Soc. Trans., v. 11, p. 138-164.

Marshall, Brian, 1967, The present status of zircon: Sedimentology, v. 9, p. 119-136.

McIntyre, D. D., 1959, The hydraulic equivalence and size distributions of some mineral grains from a beach: Jour. Geology, v. 67, p. 278-301.

Miller, D. N., Jr., and R. L. Folk, 1955, Occurrence of detrital magnetite and ilmenite in red sediments—new approach to significance of red beds: AAPG Bull., v. 39, p. 338-345.

Milner, H. B., 1962a, Sedimentary petrography: Volume 1, Methods in sedimentary petrography, 4th ed.: London, George Allen and Unwin Ltd., 643 p.

———— 1962b, Sedimentary petrography: Volume II, Principles and applications, 4th ed.: London, George Allen and Unwin Ltd., 715 p.

Pettijohn, F. J., 1941, Persistence of heavy minerals and geologic age: Jour. Geology, v. 49, p. 610-625.

Poldervaart, A., 1955, Zircon in rocks: 1. Sedimentary rocks: Am. Jour. Sci., v. 253, p. 433-461.

Rao, C. B., 1957, Beach erosion and concentration of heavy mineral sands: Jour. Sed. Petrology, v. 27, p. 143-147.

Rittenhouse, Gordon, 1943, Transportation and deposition of heavy minerals: Geol. Soc. America Bull., v. 54, p. 1725-1780.

Saxena, S. K., 1966, Evolution of zircons in sedimentary and metamorphic rocks: Sedimentology, v. 6, p. 1-34.

Simpson, G. S., 1976, Evidence of overgrowths on, and solution of, detrital garnets: Jour. Sed. Petrology, v. 46, p. 689-693.

Stanley, D. J., 1965, Heavy minerals and provenance of sand in flysch of central and southern French Alps: AAPG Bull., v. 49, p. 22-40.

Tatsumoto, Mitsunobo, and C. Patterson, 1964, Age studies of zircon and feldspar concentrates from the Franconia Sandstone: Jour. Geology, v. 72, p. 232-242.

Vistelius, A. B., 1964, Paleogeographic reconstructions by absolute age determinations of sand particles: Jour. Geology, v. 72, p. 483-486.

Vitanage, P. W., 1957, Studies of zircon types in Ceylon pre-Cambrian complex: Jour. Geology, v. 65, p. 117-138.

Wright, W. I., 1938, The composition and occurrence of garnets: Am. Mineralogist, v. 23, p. 436-449.

Clays
and
Shales

Fine-grained clastic rock and sediment nomenclature
of Ingram (1953), Folk (1961, 1968), and
Dunbar and Rodgers (1957).   from Picard (1971)

	Ingram Nomenclature No Connotation as to Breaking Characteristics	Massive	Fissile
Relative Amounts of Silt and Clay Unknown	Mudrock	Mudstone	Mud Shale
Silt > Clay	Siltrock	Siltstone	Silt Shale
Clay > Silt	Clayrock	Claystone	Clay Shale

Grain Size of Mud Fraction	*Folk Nomenclature* Soft		Indurated, Non-Fissile	Indurated, Fissile
Subequal Silt and Clay	Mud		Mudstone	Mud Shale
Over $\frac{2}{3}$ Silt	Silt		Siltstone	Silt Shale
Over $\frac{2}{3}$ Clay	Clay		Claystone	Clay Shale

	Dunbar and Rodgers Nomenclature Unconsolidated	General Term	Non-Fissile	Fissile
General Term	Mud, Dust	Mudrock, Lutite	Mudstone	Shale, Mud Shale
Particles Mainly > 4 Microns	Silt	Siltrock	Siltstone	Silt Shale
Particles Mainly < 4 Microns	Clay	Clayrock	Claystone	Clay Shale
Weakly Metamorphosed			Argillite	Clay Slate

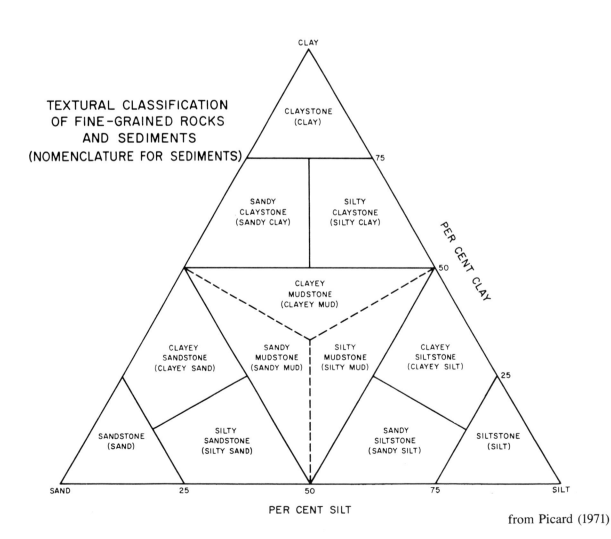

TEXTURAL CLASSIFICATION
OF FINE-GRAINED ROCKS
AND SEDIMENTS
(NOMENCLATURE FOR SEDIMENTS)

from Picard (1971)

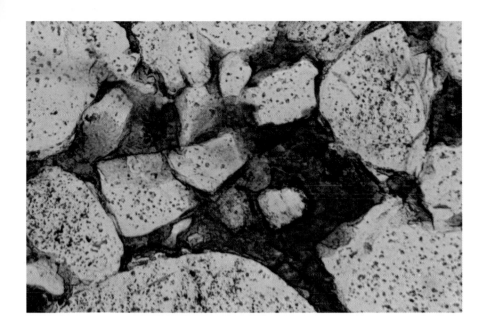

Jurassic Salt Wash Mbr. of Morrison Fm.
Colorado

Clay minerals, here seen completely filling
the pore space of a sandstone, commonly
are important constituents of clastic terrigenous sedimentary rocks. Interstitial clay
minerals can be detrital, either as originally
individual flakes or as disaggregated clay
clasts, or can be authigenic. Determining
with certainty the origin of matrix clay
minerals is difficult in many cases.

0.06 mm

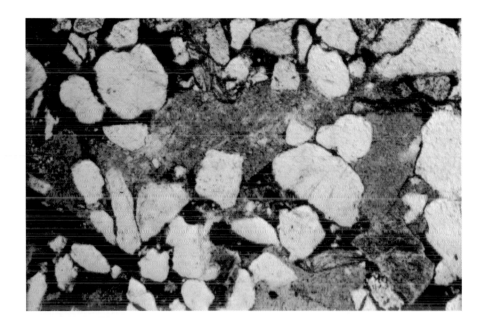

Cambrian Sillery Group
Canada (Quebec)

Interstitial clays in this example clearly
originated as shale clasts. These soft grains
were deformed during burial and flowed
around adjacent quartz grains. Diagnostic
criteria for this type of clay matrix include
clay-filled areas larger than normal pores,
remnant shale clast outlines, and patchy
distribution of clays.

0.27 mm

Devonian Oriskany Ss. ★
Virginia

A much more advanced stage of compaction of shale clasts than that in last example. Here birefringent clay minerals have
been deformed extensively and have flowed
between quartz grains to the point where it
is difficult to recognize any original clast
outlines. Note strain shadows within quartz
grains, a further indication of deformation
of the sediment.

XN   0.15 mm

66

Pennsylvanian Teuschnitzer Cgl. ★
Germany

Primary (detrital) clay matrix recognizable here by the fact that quartz grains are floating in the clay. That is, the quartz does not form a framework which could support itself, and this implies that the clays were introduced at the same time as the quartz. Cases in which primary clay was incorporated into a self-supporting grain framework are more difficult to recognize.

0.27 mm

Quaternary soil
Australia

Clays infiltrated and/or neoformed in a red podzolic soil. In areas of intense weathering clays are moved downward in the soil zone by a combination of mechanical and chemical processes (eluviation). The red color results from the accumulation of largely insoluble iron oxides and hydroxides and is common in tropical and hot arid soils. Dispersion of quartz in clay results from high "mobility" and expansive neoformation of clays. Photo by E. F. McBride.

0.30 mm

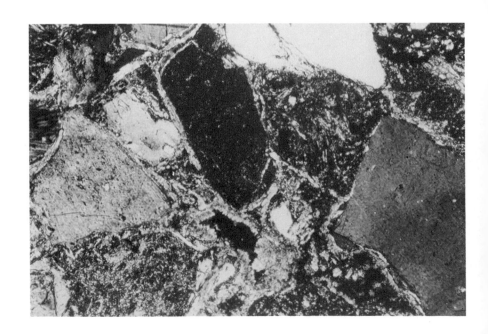

Miocene Cierbo and Neroloy Ss.
California

Authigenically formed clays (montmorillonite/smectite). The occurrence of the clay minerals as oriented, fibrous crusts lining pores in this sandstone indicates authigenic growth as a pore filling cement. Other clays appear to be forming as an alteration of unstable detrital rock fragments.

XN   0.10 mm

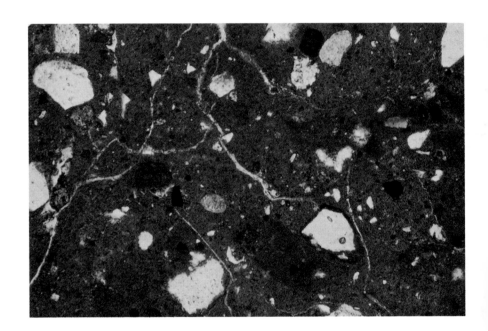

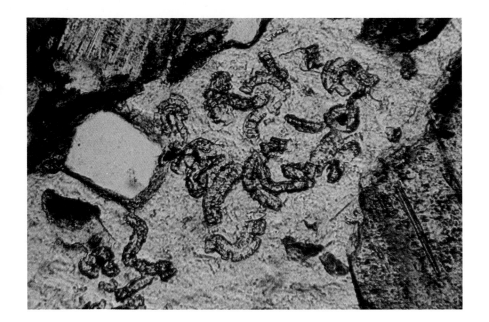

Eocene Huerfano Fm. ★
Colorado

Vermicular kaolinite as an authigenic pore-filling cement. The coarse crystal size and complex, worm-like structure of the kaolinite suggest *in situ* formation. Here, the kaolinite is embedded within a later formed crystal of pore-filling calcite cement.

0.05 mm

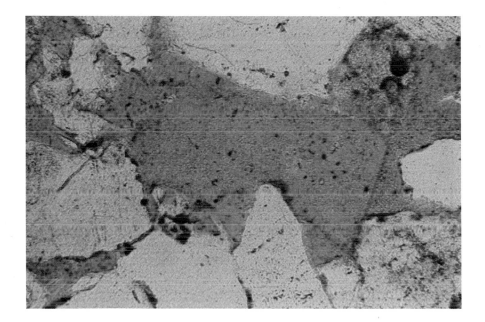

Oligocene Frio Fm.
Texas 4,826 m (15,833 ft)

Clay minerals, mainly kaolinite, partly filling both primary and secondary pore space (porosity shown with dark blue stained epoxy). Clays within leached secondary porosity probably formed as replacement of feldspar crystals which were subsequently dissolved, leaving residue of clays. Note intercrystalline porosity in clay-filled areas. Photo by R. G. Loucks.

0.11 mm

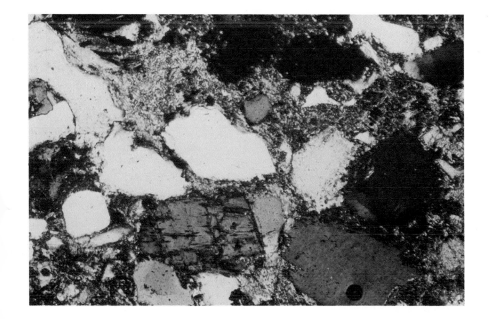

Upper Triassic New Haven Arkose
Connecticut

Complex assemblage of clay minerals including sericite, chlorite, and kaolinite. Clays were identified using both X-ray diffraction and petrographic techniques. In such complex assemblages and textures, the distinction between detrital, replacement, and pore-filling clay types becomes very difficult and uncertain.

XN   0.15 mm

Pennsylvanian Tensleep Ss.
Wyoming

Detrital matrix within sandstones is not always clay; in this example, micritic calcite matrix (carbonate mud) is present in interstices between quartz and carbonate framework grains. Micritic carbonate most commonly has a crystal size of 2-5 µm and is readily identified by its equant crystal form and extremely high birefringence.

XN   0.10 mm

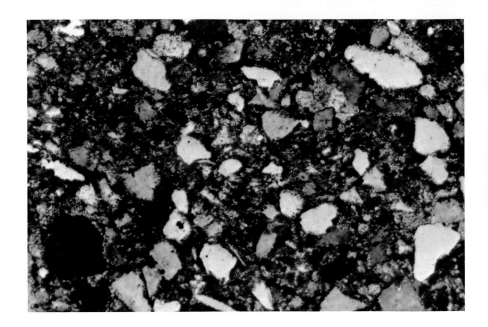

Eocene 'Plum Bentonite'
Texas

A smectite (montmorillonite) shale. Clays are best identified by X-ray diffraction techniques but can be distinguished optically to some degree. Smectite has relatively high birefringence (which differentiates it from kaolinite) and an index of refraction below epoxy, balsam, and quartz (which distinguishes it from illite and sericite). Mixed layering of smectite and illite is common and complicates identification.

XN   0.30 mm

Lower Cretaceous Mowry Shale
Wyoming

Smectitic (montmorillonitic) shale from a bentonite (Clay Spur Bentonite Bed). This SEM view shows irregular, wavy plates or sheets which are characteristic of most smectites. Photo by R. R. Larson.

SEM   1 µm

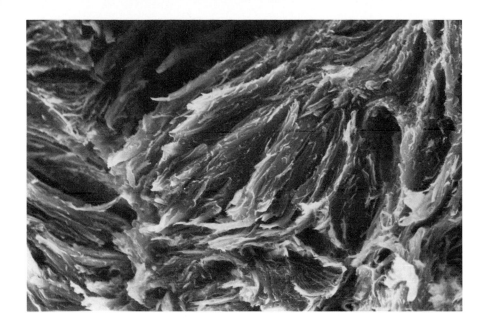

Tertiary weathering zone in basalts
Washington

An SEM view of nontronite, a member of
the smectite (montmorillonite) group of
clay minerals. Apparently formed as an
alteration of glassy material and augite
in volcanic rocks and volcaniclastic sediments. Photo by R. R. Larson.

SEM   7 μm

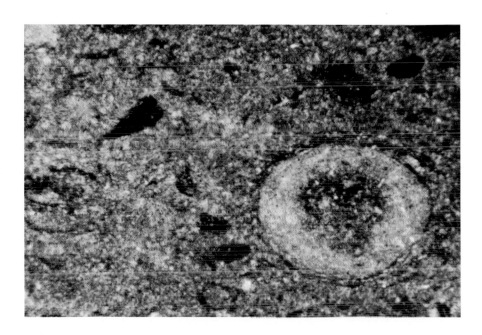

Pennsylvanian "Ball Clay"
Missouri

A kaolinite-rich claystone with abundant
ball-like structures. The origin of these
structures is not fully understood but they
may form as soil pisolites. Kaolinite is
generally the product of intense weathering
of feldspathic or clayey rocks in hot,
humid climates. In such settings potassium
is stripped from the clays, forming kaolinite.

0.30 mm

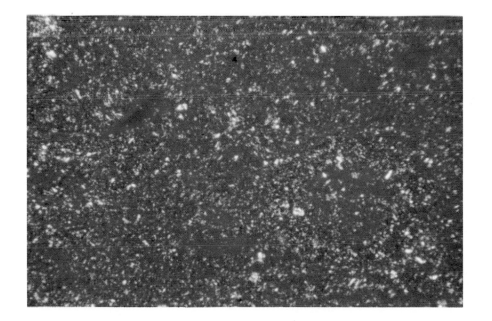

Pennsylvanian "Ball Clay"
Missouri

Same as previous photo but with crossed
polarizers. Kaolinite is distinguished from
smectite, illite, and sericite by its much
lower birefringence. However, kaolinite is
often very difficult to distinguish from
chert and finely crystalline volcanic rock
fragments. X-ray diffraction provides the
most certain method of identification.

XN   0.30 mm

Cretaceous Middendorf Fm.
South Carolina

An SEM view of a kaolinitic claystone.
Note characteristic platy structure of
stacks of kaolinite plates, perhaps slightly
etched in this example.

SEM 3 μm

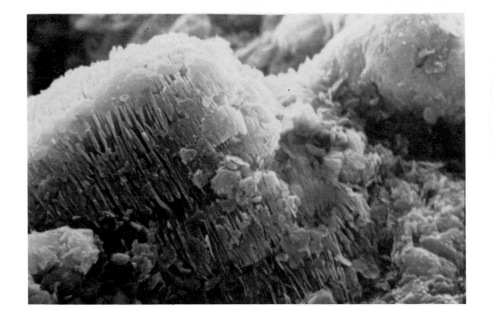

Upper Cretaceous Tuscaloosa Fm.
Louisiana 4,923 m (16,150 ft)

Authigenic kaolinite within sandstone
pores. Crystals show well developed pseudohexagonal basal scales in curved vermicular stacks, a characteristic structure for
kaolinite and related minerals such as
dickite. Photo by G. W. Smith.

SEM 6 μm

Cretaceous Black Leaf Fm.
Montana

A coarse sericite claystone. Sericite can be
distinguished, in thin section, from the
virtually identical illite only on the basis of
crystal size. Sericite and illite both have an
index of refraction higher than balsam
(which distinguishes them from smectite),
and a birefringence much higher than that
of kaolinite.

XN 0.10 mm

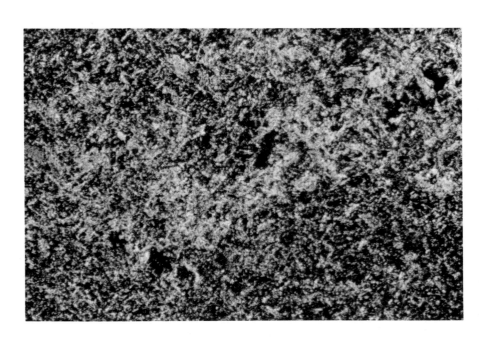

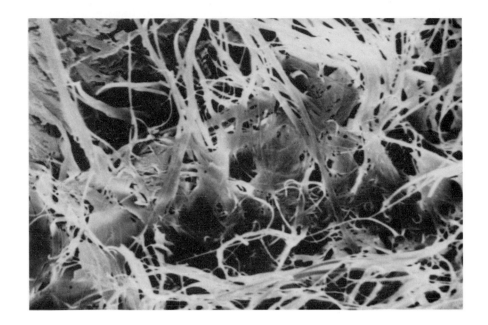

Lower Permian Rotliegendes Ss.
Leman Field, British North Sea
ca. 2,680 m (8,800 ft)

An SEM view of authigenic illite in a sandstone. In such views illite is characterized by sheet-like flakes with wispy, fibrous terminations which can commonly bridge large pores. The fibers are, in most cases, unoriented and may intertwine complexly. Photo by E. D. Pittman.

SEM   20 µm

Ordovician Griffel Schiefer ✶
Germany

Knots of authigenic chlorite (anomalous blue or "ultrablue" birefringence) have grown in a matrix of organic carbon-rich shale. The growth of chlorite in sedimentary rocks is a common phenomenon in the entire range of temperatures and pressures from moderate sedimentary burial through green-schist metamorphism. Chlorite is most readily recognized by its birefringence and morphology, but X-ray diffraction work is needed for detailed mineralogical studies.

XN   0.10 mm

Oligocene Frio Fm.
Texas 3,068 m (10,066 ft)

An SEM view of authigenic growth of knots of chlorite in pore space of a sandstone. Chlorite typically shows a platy or bladed appearance with a variety of overall habits, including the clusters or spherules shown here. Authigenic quartz can also be seen infilling pore space in this example. Photo by R. G. Loucks.

SEM   4.5 µm

Miocene Hawthorn Fm.
Florida

An SEM view of another sedimentary clay mineral, attapulgite (also termed palygorskite). This group of hydrous magnesium aluminum silicates is characterized by its distinctive rod-like or fibrous structure. It is quarried as a Fuller's earth deposit. Photo by R. R. Larson.

SEM   1 μm

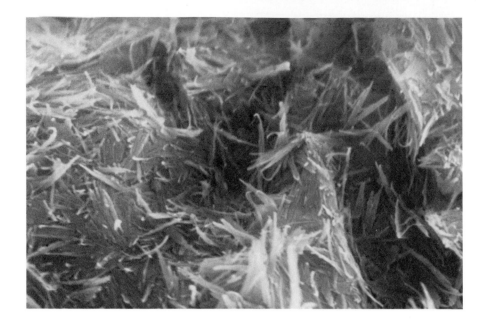

Lower Cretaceous Mowry Shale *
Utah

The classification of fine-grained sedimentary rocks generally depends on the distinction between massive and fissile types as well as a recognition of the relative proportions of clay-sized versus coarser material. This example shows extreme parallelism of clay minerals (indicated by uniform birefringence), lamination, and a marked predominance of clay-sized material. It would be termed an indurated, fissile clay shale.

XN   0.10 mm

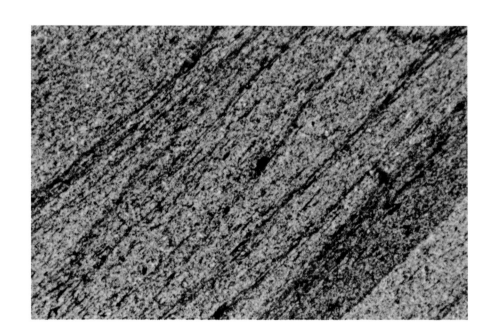

Lower Cretaceous Mowry Shale *
Utah

Same as previous photo but rotated approximately 45°. Note very apparent change in overall birefringence of sample. This indicates excellent alignment of constituent clay minerals in a strongly nonrandom (fissile) fabric.

XN   0.10 mm

Upper Cretaceous Monte Antola Fm.
Italy

An indurated nonfissile siltstone. Sample shows moderate to poor orientation of clay minerals (chlorite and sericite) and a predominance of silt- over clay-sized material.

XN   0.10 mm

Upper Cretaceous Monte Antola Fm.
Italy

A burrowed siltstone. Original lamination, if present, has been largely destroyed by churning of the sediment by burrowing organisms. In this case, burrows are clearly visible because burrow walls are packed with coarser silt whereas the burrow center is filled with finer grained, clayey material. Recognition of burrows can be important in paleoenvironmental reconstruction.

0.22 mm

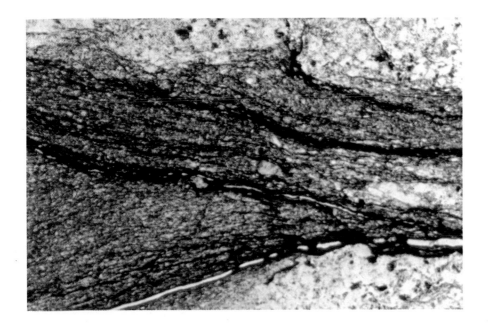

Lower Ordovician Phycoden Schichten
Germany

Solution seams in a shale. Compaction and dissolution of material has taken place in the shaly zone between two more indurated silty areas (lighter color). The solution zones or seams, also termed microstylolites or horsetail seams, have concentrations of relatively insoluble organic matter, pyrite, clay, and other minerals yielding the very dark appearance.

0.27 mm

Cretaceous-Tertiary bauxite
Arkansas

This is an example of a kaolinitic bauxite showing numerous soil-type features indicative of formation within a weathering zone. The sediment is pisolitic with individual pisolites showing irregular lamination and desiccation cracking. Calcite and iron oxides are major cements. Bauxites form in areas of very intense leaching in hot, humid climates and their recognition aids in interpretation of paleoclimates.

0.38 mm

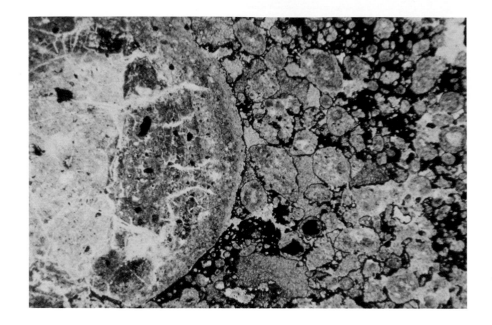

Cretaceous-Tertiary bauxite
Arkansas

Same as previous photo but with crossed polarizers. Note extremely low birefringence to virtual isotropism of bauxite and kaolinite. Reddish color is from hematite cementation. Pisolite development, as shown here, is characteristic of many soil-type deposits including bauxites and caliche crusts.

XN   0.38 mm

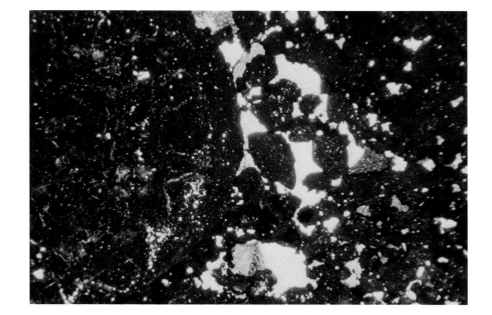

Quaternary soil *
Australia

An example of pisolitic structure in a lateritic hardpan. Sediment includes clays and quartz grains largely cemented by iron and aluminum hydroxides. Recognition of such weathering horizons can form an important component of paleogeographic reconstructions. Photo by E. F. McBride.

0.30 mm

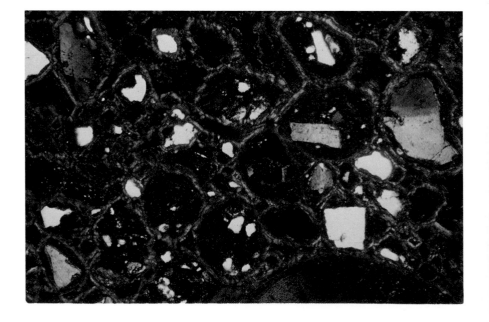

Selected Clay, Shale, and Mica Bibliography

Audley-Charles, M. G., 1967, Greywackes with a primary matrix from the Viqueque Formation (Upper Miocene-Pliocene), Timor: Jour. Sed. Petrology, v. 37, p. 5-11.

Biscaye, P. E., 1965, Mineralogy and sedimentation of Recent deep-sea clay in the Atlantic Ocean and adjacent seas and oceans: Geol. Soc. America Bull., v. 76, p. 803-832.

Borst, R. L., and W. D. Keller, 1969, Scanning electron micrographs of API reference clay minerals and other selected samples, *in* Internat. Clay Conf. Proc., Tokyo, 1969: Israel Univ. Press, v. 1, p. 871-901.

Brown, G., ed., 1961, The x-ray identification and crystal structures of clay minerals: London, Mineralogical Society, 544 p.

Caillere, S., and S. Henin, 1963, Mineralogie des Argiles: Masson et Cie., 355 p.

Carroll, Dorothy, 1970, Clay minerals: a guide to their X-ray identification: Geol. Soc. America Spec. Paper 126, 80 p.

Cody, R. D., 1971, Adsorption and the reliability of trace elements as environment indicators for shales: Jour. Sed. Petrology, v. 41, p. 461-471.

Dickinson, W. R., 1970, Interpreting detrital modes of graywacke and arkose: Jour. Sed. Petrology, v. 40, p. 695-707.

Dunbar, C. O., and John Rodgers, 1957, Principles of stratigraphy: New York, John Wiley and Sons, 356 p.

Folk, R. L., 1961, Petrology of sedimentary rocks: Austin, Texas, Hemphill's, 154 p.

—— 1968, Petrology of sedimentary rocks: Austin, Texas, Hemphill's, 170 p.

Glass, H. D., 1958, Clay mineralogy of Pennsylvanian sediments in southern Illinois: Fifth Natl. Conf. on Clays and Clay Minerals Proc., p. 227-241.

Grim, R. E., 1968, Clay mineralogy, 2nd ed.: New York, McGraw-Hill, 596 p.

Hayes, J. B., 1970, Polytypism of chlorite in sedimentary rocks: Clays and Clay Minerals, v. 18, p. 285-306.

Ingram, R. L., 1953, Fissility in mudrocks: Geol. Soc. America Bull., v. 65, p. 869-878.

Jonas, E. C., 1961, Mineralogy of the micaceous clay minerals: 21st Internat. Geol. Cong., Proc., pt. 24, p. 7-16.

Keller, W. D., 1956, Clay minerals as influenced by environments of their formation: AAPG Bull., v. 40, p. 2689-2710.

—— 1970, Environmental aspects of clay minerals: Jour. Sed. Petrology, v. 40, p. 788-813.

—— 1977, Scan electron micrographs of kaolins collected from diverse environments of origin—IV. Georgia kaolin and kaolinizing source rocks: Clays and Clay Minerals, v. 25, p. 311-345.

Kossavskaya, A. G., and V. A. Drits, 1970, The variability of micaceous minerals in sedimentary rocks: Sedimentology, v. 15, p. 83-101.

Leibling, R. S., and H. S. Scherp, 1976, Chlorite and mica as indicators of depositional environment and provenance: Geol. Soc. America Bull., v. 87, p. 513-514.

Mackenzie, R. C., and B. D. Mitchell, 1966, Clay mineralogy: Earth-Sci. Rev., v. 2, p. 47-91.

Mansfield, C. F., and S. W. Bailey, 1972, Twin and pseudotwin intergrowths in kaolinite: Am. Mineralogist, v. 57, p. 411-425.

Millot, G., 1970, Geology of clays (translated by W. R. Forrand and H. Paquet): New York, Springer-Verlag, 429 p.

—— J. Lucas, and R. Wey, 1963, Research on evolution of clay minerals and argillaceous and siliceous neoformation, *in* Clays and Clay Minerals, 10th National Conf.: New York, Pergamon Press, p. 399-412.

O'Brien, N. R., 1970, The fabric of shale—an electron-microscope study: Sedimentology, v. 15, p. 229-246.

Parham, W. E., 1966, Lateral variations of clay mineral assemblages in modern and ancient sediments: Internat. Clay Conf. Proc., Jerusalem, v. 1, p. 135-145.

Picard, M. D., 1971, Classification of fine-grained sedimentary rocks: Jour. Sed. Petrology, v. 41, p. 179-195.

Reynolds, R. C., and J. Hower, 1970, The nature of interlayering in mixed layer illite-montmorillonites: Clays and Clay Minerals, v. 18, p. 25-36.

Smoot, T. W., 1960, Clay mineralogy of pre-Pennsylvanian sandstones and shales of the Illinois Basin, Part III—Clay minerals of various facies of some Chester Formations: Illinois Geol. Survey Circ. 293, p. 1-19.

Weaver, C. E., 1958, Geologic interpretation of argillaceous sediments, Part I. Origin and significance of clay minerals in sedimentary rocks: AAPG Bull., v. 42, p. 254-271.

—— 1959, The clay petrology of sediments: 6th National Conf. on Clays and Clay Minerals, Proc., p. 154-187.

—— and L. D. Pollard, 1973, The chemistry of clay minerals: New York, Elsevier.

Associated Sediments

Cretaceous Travis Peak Fm.
Texas

A detrital chert fragment. Although this grain shows a uniform, dark, microcrystalline quartz fabric with no relict depositional features, many other cherts (both as detrital grains and *in situ* deposits) contain an abundance of preserved details which may shed important light on paleoenvironmental conditions. Note fine crystal size and low birefringence of chert.

XN   0.10 mm

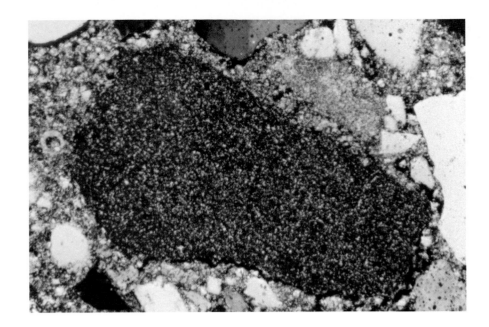

Upper Cretaceous White Limestone
Northern Ireland

A chert nodule formed as a replacement of a fine-grained limestone. Abundant carbonate inclusions are preserved within the chert. Areas of originally more coarsely crystalline carbonate (such as the Foraminifera in the center of photo) have been replaced by more coarsely crystalline quartz. This type of fabric-controlled replacement is common in cherts.

XN   0.10 mm

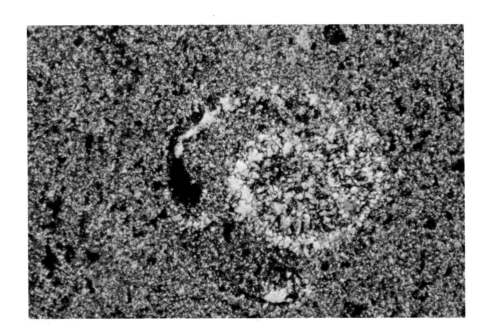

Upper Cretaceous Upper Chalk ★
England

A spicular chert. Abundant monaxon and triaxon sponge spicules preserved as megaquartz in a matrix of microquartz (chert) of replacement origin. Sponge spicules are frequently preserved in cherts, in part because they are commonly the source of the replacement silica. Dissolution of opal (either from biogenic or volcanic sources) provides most of the silica for replacement chertification.

XN   0.27 mm

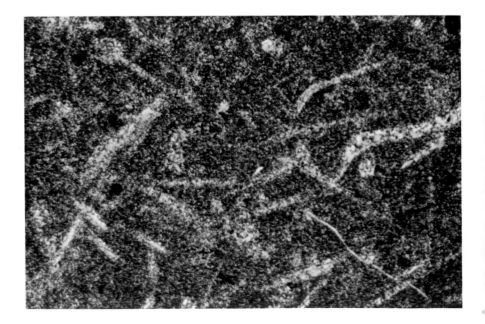

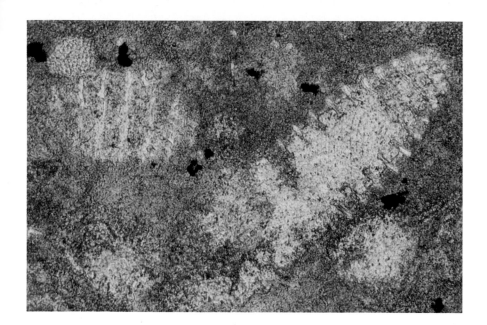

Upper Jurassic Point Sal ophiolite group
California

Cross sections through several Radiolaria
showing moderately good preservation of
the tests despite alteration from original
opal to present microquartz chert. Radiolaria are another major source of replacement silica in many sections. Radiolarian
remains can be recognized on the basis of
the coarse pore structure as well as the size
and shape of the tests.

0.07 mm

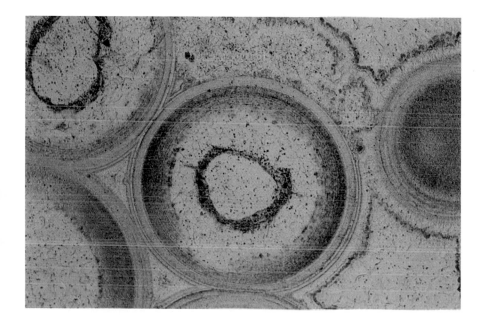

Upper Cambrian Mines Dolomite Mbr. of
  Gatesburg Fm.
Pennsylvania

Silica replacement of an oolitic limestone.
This is an example of the excellent fidelity
of preservation which is possible with
chert-replacement fabrics. Brownish color
is a result of numerous very small inclusions of water- and air-filled vacuoles
within the chert structure and is typical of
most chert.

0.27 mm

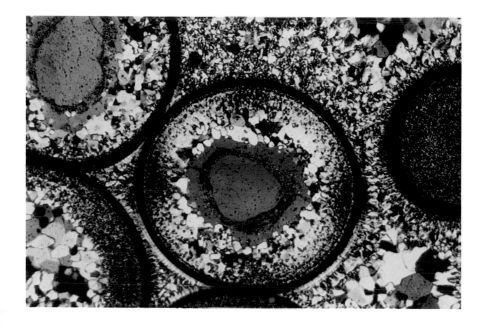

Upper Cambrian Mines Dolomite Mbr. of
  Gatesburg Fm. ★
Pennsylvania

Same as previous view but with crossed
polarizers. Shows the broad range of silica
fabrics which can be found even in a small
area. Detrital quartz grains (oolitic nuclei)
have overgrowths. Oolitic carbonate was
replaced by micro- and megaquartz whereas
the intergranular pore space was filled with
a lining of chalcedony followed by megaquartz.

XN   0.27 mm

Cretaceous chert pebble in Holocene
  sediments
Texas

Silica replacement of limestone. Shows the
three main forms of sedimentary quartz—
chert or microquartz, chalcedony, and
megaquartz. Chert here replaced the bulk
of the fine-grained carbonate sediment.
The radial fibrous chalcedony lined a
former cavity which was subsequently
completely infilled with equant megaquartz. This section is slightly thicker than
the standard 30 μm.

XN   0.38 mm

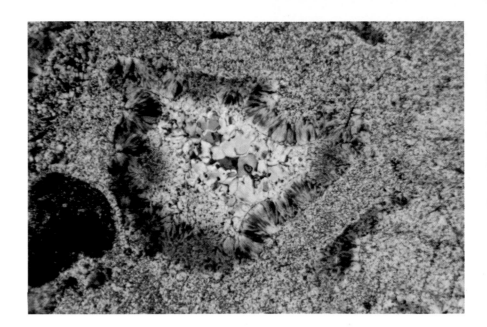

Tertiary 'Vieja Group' ⋆
Texas

Chert filling of a large void between breccia
fragments in a volcaniclastic rock. The void
was first lined with several generations of
chalcedony separated by hematitic bands.
Final filling of the void was accomplished
by a sequence of graded chert bands.
Although the origin of this type of graded
chert filling is unclear, the texture is not
uncommon.

XN   0.38 mm

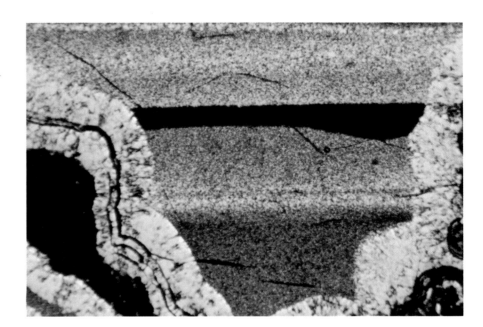

Middle Paleozoic Caballos Novaculite
Texas

"Cubic" or "fortification zoned" quartz.
These cavity lining crystals in jasper beds
have cubic outlines and zonation of brown
(fluid-inclusion-rich) and white layers.
Such quartz morphology has been proposed as a criterion for recognizing the
association of cherts with ancient evaporites
(McBride and Folk, 1977).

                0.27 mm

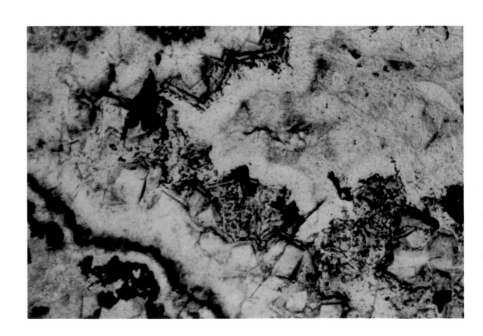

Upper Jurassic Radiolariti
Italy

Chalcedonic quartz infilling of a cavity. Commonly, as in this example, chalcedony occurs as radiating bundles of fibers. In this case, the bundles increase in size from the margins of the cavity to the center. Chalcedony, like most other forms of microquartz, contains numerous micro-inclusions filled with liquid. These inclusions are responsible for the relatively low refractive index of chalcedony.

XN   0.27 mm

Upper Jurassic Radiolariti ★
Italy

"Zebraic" chalcedony. This is a fibrous microquartz cavity lining in which the fibers are alternately light and dark as one views along the fiber elongation direction under polarized light. McBride and Folk (1977) have described the association of zebraic chalcedony with evaporite minerals.

XN   0.27 mm

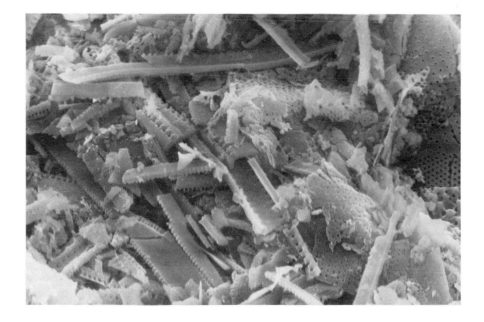

Miocene-Pliocene Monterey Shale
California

A diatomite (a primary siliceous sediment). This sample contains abundant, largely unaltered fragments of marine diatoms still preserved as original opaline silica. Most extensive bedded cherts of Phanerozoic age probably originated from similar deposits of biogenic silica (Radiolaria, diatoms, or siliceous sponges).

SEM   2 μm

82

Upper Cretaceous Craie grise
Netherlands

Spherules (sometimes called lepispheres) of cristobalite. Recent work has shown that the alteration of biogenic opal to quartz chert generally goes through the intermediate stage of cristobalite. The transitions between these three phases are largely temperature controlled although time and associated lithologies also play a large role. In general, the presence of clay minerals retards the transformation; $CaCO_3$ accelerates it.

SEM   2.6 μm

Mississippian Redwall Ls.
Arizona

Electron micrograph of a carbon replica of bedded chert. Sample shows equant crystals which range in size from very fine to medium crystalline microquartz. The various forms of sedimentary silica (biogenic opal, cristobalite, and alpha quartz) are generally distinct and recognizable using the electron microscope. Photo by J. C. Hathaway, courtesy of E. D. McKee.

TEM   3.5 μm

Pliocene Bone Valley Fm.
Florida

Detrital phosphate grains (cloudy) associated with detrital quartz (clear). The phosphatic grains, which are composed of several phosphate minerals, generally lack internal structure in this example although grains with oolitic or pisolitic structure are common in other deposits. The phosphate in this case was formed in marine conditions (associated teeth and bones of marine animals) but was concentrated by weathering and erosion.

0.38 mm

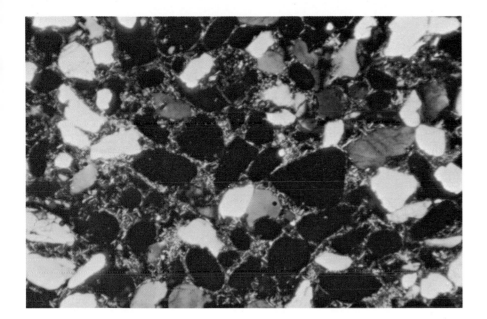

Pliocene Bone Valley Fm.
Florida

Same as previous photo but with crossed polarizers. Phosphate (primarily fluorapatite) here is virtually isotropic. The gray-to-brown color in transmitted light and extremely low birefringence in cross-polarized light characterize most phosphate minerals. Staining and X-ray diffraction are also useful in the identification of phosphates.

XN   0.38 mm

Permian Phosphoria Fm. ★
Idaho

An oolitic phosphate deposit. Nuclei of ooids are phosphatic bone fragments. These are coated with one or more layers of apatitic phosphate. The continuity of coating layers implies mobility of the grains probably due to wave or current agitation. Phosphate accumulations are often associated with nondepositional intervals or periods of slow sedimentation, and also can imply high biological productivity.

0.38 mm

Tertiary 'Vieja Group'
Texas

Section of a piece of phosphatic bone material (collophane) with complex but typical internal structure. The patterns are created by phosphatic material surrounding small channels within the bone. Note the very low order interference colors.

XN   0.38 mm

Upper Cretaceous Atco Fm.
 (Austin Group)
Texas

A phosphatic bone fragment as seen in
SEM. Such grains are normally identifiable
on the basis of their surface texture (as
with the dimpled pattern seen here) or by
their porous internal structure when seen
in cross section.

SEM   100 μm

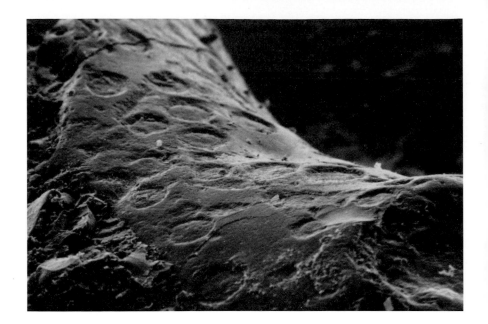

Mississippian upper Debolt Fm.
Canada

The recognition of evaporite minerals can
be important in the reconstruction of the
geologic history of sedimentary basins. In
this example, crystals of probable celestite
have formed within a micritic carbonate
matrix. Note the low birefringence and
crystal shapes similar to gypsum. Several
crystals have been replaced by calcite (high
birefringence).

XN   0.25 mm

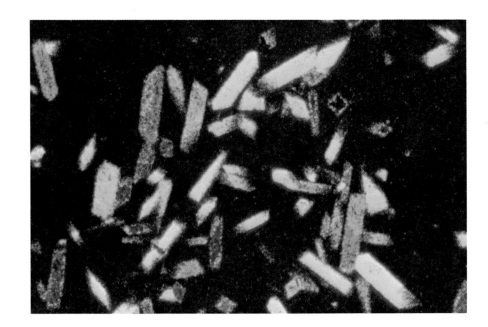

Upper Permian Castile Fm.
New Mexico

A bedded and varved gypsum deposit. In
the upper part of the photo is a lamina of
orange-brown calcite crystals whereas the
rest of the slide shows interlocking crystals
and crystal fragments of gypsum. In the
deeper subsurface this rock is anhydrite
and it probably has inverted to gypsum
during uplift and erosion. Note characteristic low birefringence of gypsum.

XN   0.27 mm

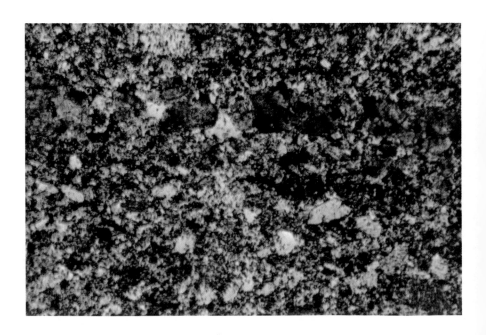

Upper Miocene (Messinian) Gesso Solfifera
Italy

Detrital gypsum crystals and carbonate matrix reworked by turbidity currents. Note low birefringence, crystal outlines, and presence of twinning in many of the gypsum crystals and crystal fragments. Although quite soluble and soft, gypsum can, under some circumstances be transported significant distances and be preserved even in deeper water sediments.

XN   0.27 mm

Holocene dunes
Utah

Detrital (windblown) gypsum crystals in modern dunes flanking the Bonneville Salt Flats. In arid climates and with relatively short distance transport, gypsum crystals can remain well preserved. Characteristic discoidal shapes, carbonate inclusions, etching, and low birefringence can be seen in this example.

XN   0.27 mm

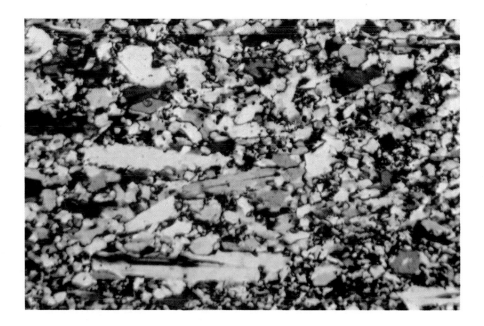

Upper Miocene (Messinian) unit
Cyprus

A common form of occurrence of gypsum —as lath-like crystals. Twinning can be seen in some of the crystals. Low birefringence, closely interlocking fabric, and relative scarcity of inclusions are visible in this example.

XN   0.10 mm

Upper Miocene (Messinian) Gesso Solfifera
Italy

On occasion, gypsum can form oolitic
growths as seen here. Nucleus is a rounded
gypsum crystal fragment which is coated
with numerous concentric (and slightly
eccentric) layers of gypsum marked by
brown bands of inclusions. Ooids are
incorporated in a groundmass of gypsum
cement. These grains formed in agitated
shallow marine water.

0.27 mm

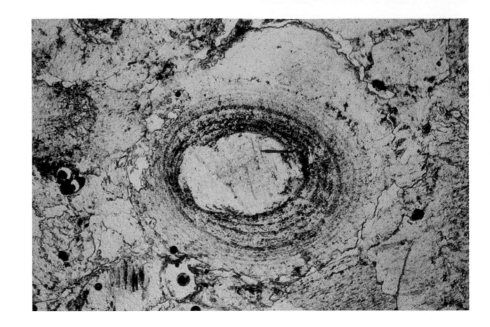

Upper Miocene (Messinian) Gesso Solfifera
Italy

Same as previous photo but with crossed
polarizers. Although ooids are composed of
thin, concentric laminations, the crystal
fabric cuts across this texture. Crystals of
gypsum are arranged in a radial pattern
which extends from the outer margin of
the nucleus to the edge of the ooid. This is
analogous to fabrics found in carbonate
ooids formed in hypersaline settings.

XN   0.27 mm

Upper Miocene (Messinian) Eraclea Minoa
    mass flow unit
Italy (Sicily)

Gypsum cement in a *Globigerina* ooze.
Large gypsum cement crystals, derived
largely from detrital gypsum within the
unit, fill virtually all porosity in this sediment. Early pore filling by chemically
unstable cement provides the potential for
later diagenetic removal of the cement and
the generation of large volumes of secondary porosity.

XN   0.10 mm

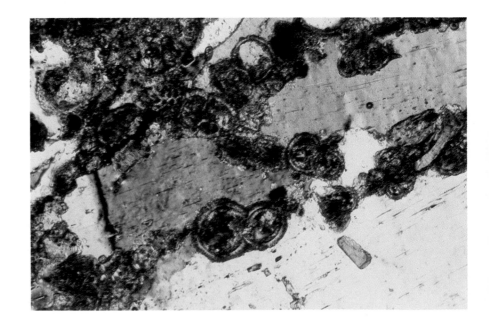

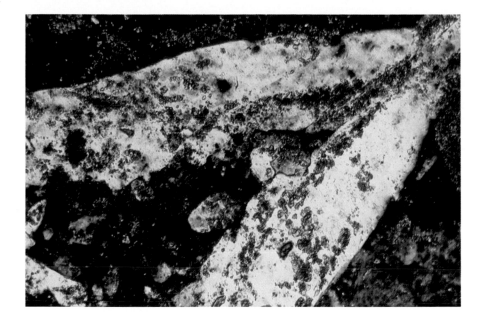

Pleistocene Montallegro Fm.
Italy (Sicily)

Large gypsum crystals which have grown within the enclosing sediments. Note the abundant carbonate inclusions within the crystal, arranged parallel with crystallo graphic directions. This is a very common phenomenon in rapidly grown evaporite crystals.

XN   0.10 mm

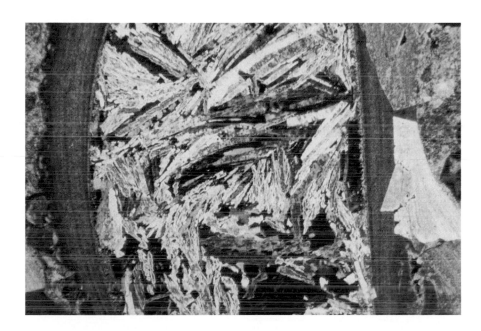

Lower Cretaceous Ferry Lake Anhydrite
Texas

Fibrous to bladed anhydrite crystals fill the central cavity of a serpulid worm tube. Coarser anhydrite has replaced parts of the serpulid tube walls. Anhydrite is readily identifiable on the basis of its moderately high birefringence and its crystal shapes.

XN   0.38 mm

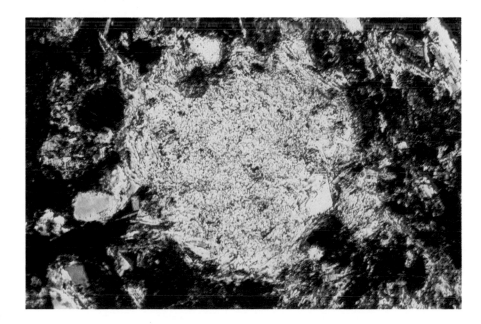

Upper Miocene (Messinian) unit
Mediterranean (DSDP Leg 13)

A replacive and/or displacive anhydrite nodule in foraminiferal sediments. The anhydrite consists of numerous small cleavage fragment produced during displacive growth of the nodule. Though these types of nodules are common in sabkha environments, they are by no means restricted to such settings.

XN   0.27 mm

Silurian Salina Fm. ★
Michigan

Anhydrite laths and calcitic fossil fragments cemented by halite. Halite is recognizable in this view only by the presence of
thin cleavage traces through what otherwise
looks like void space. The lack of compactional deformation of this very loosely
packed rock implies very early formation
of anhydrite and halite.

0.27 mm

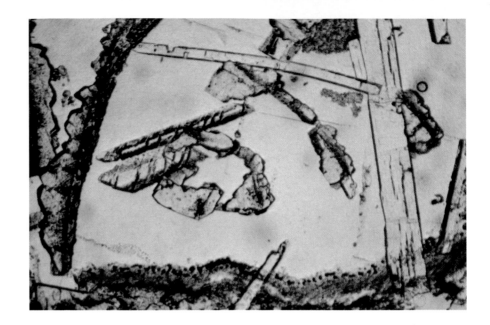

Silurian Salina Fm. ★
Michigan

Same as previous photo but with crossed
polarizers. Halite is completely isotropic
and thus is very difficult to recognize in
polarized light. Anhydrite laths have
moderately high birefringence and can be
seen to fill void space and partially replace
calcitic fossils. Calcite has extremely high
birefringence and occurs both in thin-
shelled fossils and as cement crystals.

XN   0.27 mm

Upper Miocene (Messinian) unit
Mediterranean (DSDP Leg 13)

A halite bed composed of hopper crystals.
The hoppers illustrate the cubic habit of
halite and show extensive zonation of
inclusion rich and inclusion-poor layers.
These crystals, as with the halite in previous
photos, are completely isotropic in cross-
polarized light.

0.27 mm

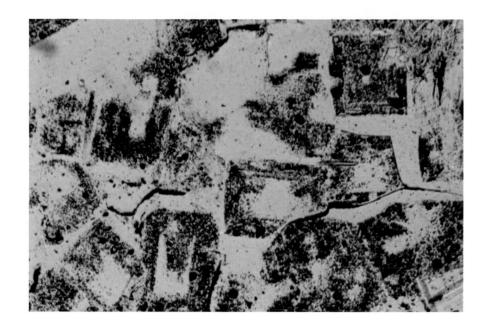

Upper Miocene (Messinian) unit *
Mediterranean (DSDP Leg 13)

Detail of halite hopper bed in previous photo. Zonation of fluid (and scarcer mineral) inclusions can be clearly seen, as can the cubic structure of the mineral. In the absence of inclusions, halite can be easily overlooked in thin section. Furthermore, special precautions (such as grinding in oil) should be taken during section preparation when halite is expected.

0.10 mm

Upper Permian Salado Fm.
New Mexico

Sylvite (KCl) has a very similar crystallographic structure to halite (both are isometric hexoctahedral with cubic cleavage). Both minerals also have isotropic behavior under cross-polarized light. Sylvite has a lower index of refraction than halite and frequently contains inclusions which result in color tinting of the crystal, as with the reddish-brown hues in this example.

0.27 mm

Pliocene Big Sandy Fm.
Arizona

Tuffaceous intervals are frequently encountered in sedimentary sections. This photo shows a tuff in which smectitic clays outline the volcanic shards. The shards themselves, although originally composed of volcanic glass, have been replaced by clinoptilolite, a zeolite. In spite of this extensive alteration, original shard texture is still readily visible.

0.10 mm

Tertiary Horse Springs Fm.
Nevada

Volcanic glass shards and other minerals derived from a volcanic source in a calcite-cemented sandstone. The shards look slightly brown-yellow when compared with the detrital quartz and feldspar in this example, largely because of the large amounts of water incorporated into the volcanic glass. Note the very angular, fragile-looking shapes characteristic of shard textures.

0.24 mm

Tertiary Horse Springs Fm.
Nevada

Same as previous view but with crossed polarizers. All volcanic glass shards are black, illustrating the isotropic behavior of glass. Calcite cement can be seen to be poikilotopic with single crystals encompassing numerous detrital grains. Quartz, plagioclase, biotite, and heavy minerals of volcanic derivation also can be seen.

XN   0.24 mm

Pleistocene Yellowstone Group (tuff) ★
Wyoming

Where tuffaceous material is deposited while still hot (as in ignimbrites or welded tuffs) the glassy shards may fuse together and be distorted by flowage, as in this example.

0.10 mm

Selected Associated Sediments Bibliography

Silica

Bissell, H. J., 1959, Silica in sediments of the Upper Paleozoic of the Cordilleran area, in H. A. Ireland, ed., Silica in sediments: Soc. Econ. Paleontologists and Mineralogists Spec. Pub. No. 7, p. 150-185.

Eugster, H. P., 1967, Hydrous sodium silicates from Lake Magadi, Kenya: Precursors of bedded chert: Science, v. 157, p. 1177-1180.

Folk, R. L., and C. E. Weaver, 1952, A study of the texture and composition of chert: Am. Jour. Sci., v. 250, p. 498-510.

Goldstein, A., Jr., 1959, Cherts and novaculites of Ouachita facies, in H. A. Ireland, ed., Silica in sediments: Soc. Econ. Paleontologists and Mineralogists Spec. Pub. No. 7, p. 135-149.

Jonas, J. B., and E. R. Seguit, 1971, The nature of opal I: Nomenclature and constituent phases: Jour. Geol. Soc. Australia, v. 18, p. 57-68.

Midgley, H. G., 1951, Chalcedony and flint: Geol. Mag., v. 88, p. 179-184.

Peterson, M. N. A., and C. C. von der Borch, 1965, Chert: Modern inorganic deposition in a carbonate-precipitating locality: Science, v. 149, p. 1501-1503.

Pittman, J. S., Jr., 1959, Silica in Edwards Limestone, Travis County, Texas, in H. A. Ireland, ed., Silica in sediments: Soc. Econ. Paleontologists and Mineralogists Spec. Pub. No. 7, p. 121-134.

Shepherd, Walter, 1972, Flint: Its origin, properties, and uses: London, Faber and Faber, 255 p.

Smith, W. E., 1960, The siliceous constituents of chert: Geol. en Mijnbouw, v. 22, new series, p. 1-8.

Swineford, A., and P. C. Franks, 1959, Opal in the Ogallala Formation in Kansas, in H. A. Ireland, ed., Silica in sediments: Soc. Econ. Paleontologists and Mineralogists Spec. Pub. No. 7, p. 111-120.

White, J. F., and T. F. Corwin, 1960, Synthesis and origin of chalcedony: Am. Mineralogist, v. 46, p. 112-119.

Phosphates

Baturin, G. N., 1971, The stages of phosphorite formation on the ocean floor: Nature Phys. Sci., v. 232, p. 61-62.

Bromley, R. G., 1967, Marine phosphorites as depth indicators: Marine Geology, v. 5, p. 503-509.

Bushinsky, G. I., 1964, On shallow water origin of phosphorite sediments, in L. M. J. U. Straaten, ed., Deltaic and shallow marine sediments; Developments in Sedimentology 1: New York, Elsevier, p. 62-70.

Cressman, E. R., and R. W. Swanson, 1964, Stratigraphy and petrology of the Permian rocks of southwestern Montana: U.S. Geol. Survey Prof. Paper 313-C, p. 275-569.

D'Anglejan, B. F., 1967, Origin of marine phosphorites off Baja California, Mexico: Marine Geology, v. 5, p. 15-44.

Kramer, J. R., 1964, Sea water: Saturation with apatites and carbonates: Science, v. 146, p. 637-638.

McKelvey, V. E., 1967, Phosphate deposits: U.S. Geol. Survey Bull. 1252-D, 21 p.

Youssef, M. I., 1965, Genesis of bedded phosphates: Econ. Geology, v. 60, p. 590-600.

Evaporites

Anderson, R. Y., W. E. Dean, Jr., D. W. Kirkland, and H. I. Snider, 1972, Permian Castile varved evaporite sequence, west Texas and New Mexico: Geol. Soc. America Bull., v. 83, p. 59-86.

Borchert, Hermann, and R. O. Muir, 1964, Salt deposits, the origin, metamorphism and deformation of evaporites: London, D. Van Nostrand Co., 338 p.

Briggs, J. I., Jr., 1963, Petrography of salt, in V. W. Kaufman, ed., Sodium chloride: New York, Reinhold Publ. Co., p. 22-27.

Butler, G. P., 1969, Modern evaporite deposition and geochemistry of coexisting brines; the sabkhas, Trucial Coast, Arabian Gulf: Jour. Sed. Petrology, v. 39, p. 79-80.

—— 1970, Secondary anhydrite from a sabkha, northwest Gulf of California, Mexico, in J. L. Rau, and L. F. Dellwig, eds., Third symposium on salt: Cleveland, Ohio, Northern Ohio Geol. Soc., v. 1, p. 153-155.

Dean, W. E., G. R. Davies, and R. Y. Anderson, 1975, Sedimentological significance of nodular and laminated anhydrite: Geology, v. 3, p. 367-372.

Edinger, S. W., 1973, The growth of gypsum: Jour. Crystal Growth, v. 18, p. 217-224.

Fuller, J. G. C. M., and J. W. Porter, 1969, Evaporite formation with petroleum reservoirs in Devonian and Mississippian of Alberta, Saskatchewan, and North Dakota: AAPG Bull., v. 53, p. 902-926.

Ham, W. E., 1962, Economic geology and petrology of gypsum and anhydrite in Baline County, in Geology and mineral resources of Blaine County, Oklahoma: Okla. Geol. Survey Bull. 89, p. 100-151.

Hardie, L. A., and H. P. Eugster, 1971, The depositional environment of marine evaporites: A case for shallow clastic accumulation: Sedimentology, v. 16, p. 187-220.

Holliday, D. W., 1970, The petrology of secondary gypsum rocks: a review: Jour. Sed. Petrology, v. 40, p. 734-744.

Hsü, K. J., 1972, Origin of saline grants—A critical review after discovery of the Mediterranean evaporite: Earth-Sci. Reviews, v. 8, p. 371-396.

Kinsman, D. J. J., 1966, Gypsum and anhydrite of Recent age, Trucial Coast, Persian Gulf, in Second symposium on salt: Cleveland, Ohio, Northern Ohio Geol. Soc., v. 1, p. 302-326.

—— 1969, Modes of formation, sedimentary associations, and diagnostic features of shallow-water and supratidal evaporites: AAPG Bull., v. 53, p. 830-840.

—— 1976, Evaporites: Relative humidity control of primary mineral facies: Jour. Sed. Petrology, v. 46, p. 273-279.

Murray, R. C., 1964, Origin and diagenesis of gypsum and anhydrite: Jour. Sed. Petrology, v. 34, p. 512-523.

Peterson, J. A., and R. J. Hite, 1969, Pennsylvanian evaporite-carbonate cycles and their relation to petroleum occurrence, southern Rocky Mountains: AAPG Bull., v. 53, p. 884-908.

Ramsdell, L. S., and E. P. Partridge, 1929, The crystal forms of calcium sulfate: Am. Mineralogist, v. 14, p. 59-74.

Raup, O. B., 1970, Brine mixing: An additional mechanism for formation of basin evaporites: AAPG Bull., v. 54, p. 2246-2259.

Schlager, W., and H. Bolz, 1977, Clastic accumulation of sulphate evaporites in deep water: Jour. Sed. Petrology, v. 47, p. 600-609.

Schreiber, B. C., G. M. Griedman, A. Decima, and E. Schreiber, 1976, Depositional environments of upper Miocene (Messinian) evaporite deposits of the Sicilian Basin: Sedimentology, v. 23, p. 729-760.

Schwerdtner, W. M., 1964, Genesis of potash rocks in Middle Devonian Prairie Evaporite Formation of Saskatchewan: AAPG Bull., v. 48, p. 1108-1115.

Shearman, D. J., 1966, Origin of marine evaporites by diagenesis: Inst. Mining Met. Trans., Sec. B., v. 75, p. 208-215.

—— and J. G. Fuller, 1969, Anhydrite diagenesis, calcitization and organic laminates, Winnipegosis Formation, Middle Devonian, Saskatchewan: Canad. Petrol. Geol. Bull., v. 17, p. 496-525.

—— 1970, Recent halite rock, Baja California, Mexico: Inst. Mining Met. Trans., v. B79, p. 155-162.

Tuffs

Ross, C. S., and R. L. Smith, 1961, Ash-flow tuffs: their origin, geologic relations, and identification: U.S. Geol. Survey Prof. Paper 366, 81 p.

Sandstone Classification

MAJOR SANDSTONE CLASSIFICATIONS

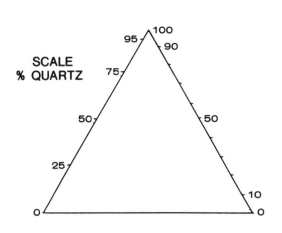

SCALE
% QUARTZ

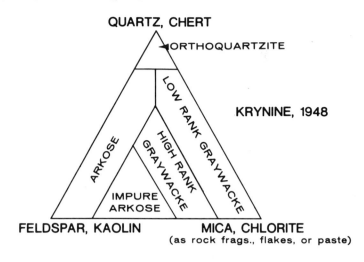

KRYNINE, 1948

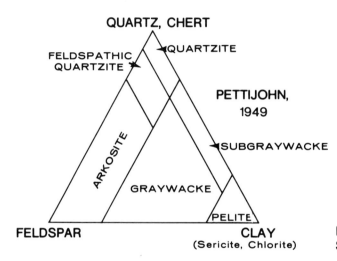

PETTIJOHN, 1949

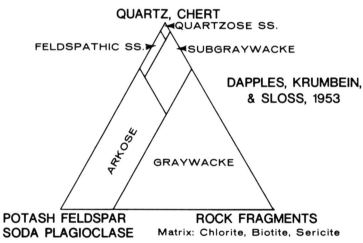

DAPPLES, KRUMBEIN, & SLOSS, 1953

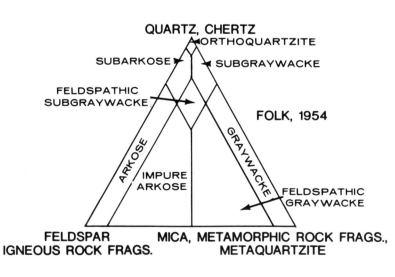

FOLK, 1954

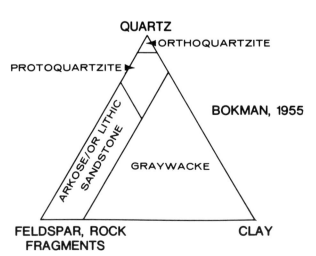

BOKMAN, 1955

QUARTZ,CHERT

SHALLOW WATER

QUARTZOSE SANDSTONE

FELDSPATHIC SANDSTONE and
SUBLABILE SANDSTONE

ARKOSE and
LABILE SANDSTONE

PELITE

UNSTABLE MINERALS,
ROCK FRAGMENTS

MATRIX
Fine silt and clay

PACKHAM,
1954

QUARTZ,CHERT

SUBGRAYWACKE

SUBLABILE GRAYWACKE

DEEP WATER

LABILE GRAYWACKE

PELITE

UNSTABLE MINERALS,
ROCK FRAGMENTS

MATRIX
Fine silt and clay

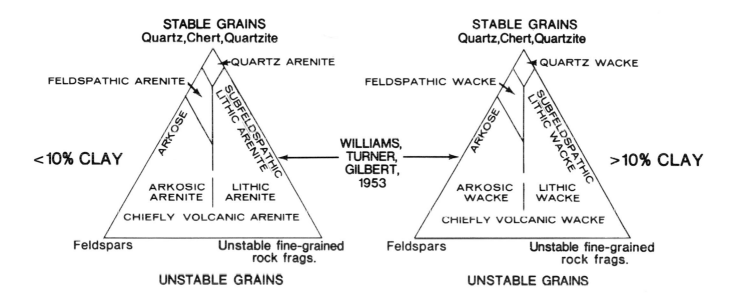

STABLE GRAINS
Quartz,Chert,Quartzite

QUARTZ ARENITE

FELDSPATHIC ARENITE

ARKOSE

SUBFELDSPATHIC
LITHIC ARENITE

ARKOSIC
ARENITE

LITHIC
ARENITE

CHIEFLY VOLCANIC ARENITE

Feldspars

Unstable fine-grained
rock frags.

UNSTABLE GRAINS

<10% CLAY

WILLIAMS,
TURNER,
GILBERT,
1953

>10% CLAY

STABLE GRAINS
Quartz,Chert,Quartzite

QUARTZ WACKE

FELDSPATHIC WACKE

ARKOSE

SUBFELDSPATHIC
LITHIC WACKE

ARKOSIC
WACKE

LITHIC
WACKE

CHIEFLY VOLCANIC WACKE

Feldspars

Unstable fine-grained
rock frags.

UNSTABLE GRAINS

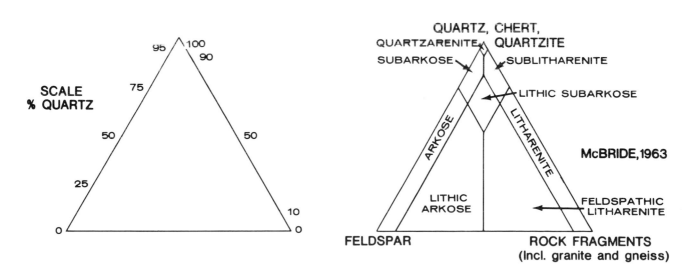

SCALE
% QUARTZ

100
95
90
75
50
50
25
10
0
0

QUARTZ, CHERT,
QUARTZITE

QUARTZARENITE

SUBARKOSE

SUBLITHARENITE

LITHIC SUBARKOSE

ARKOSE

LITHARENITE

McBRIDE,1963

LITHIC
ARKOSE

FELDSPATHIC
LITHARENITE

FELDSPAR

ROCK FRAGMENTS
(Incl. granite and gneiss)

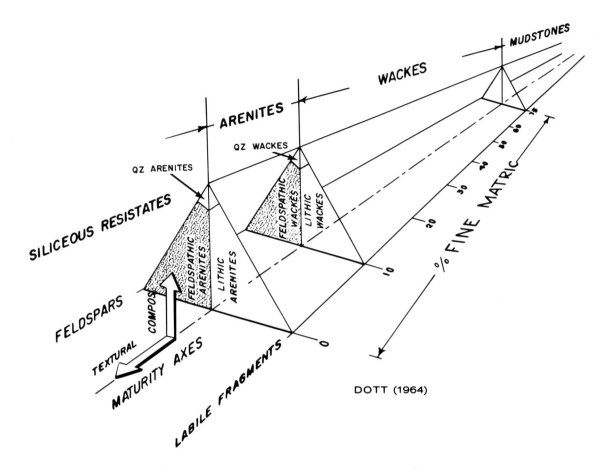

SILICEOUS RESISTATES

FELDSPARS

TEXTURAL COMPOS

MATURITY AXES

LABILE FRAGMENTS

QZ ARENITES

QZ WACKES

FELDSPATHIC ARENITES

LITHIC ARENITES

FELDSPATHIC WACKES

LITHIC WACKES

ARENITES

WACKES

MUDSTONES

% FINE MATRIX

DOTT (1964)

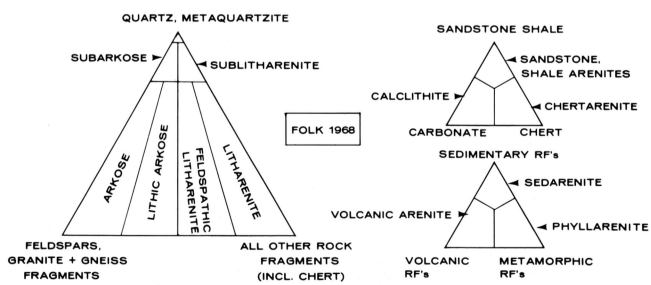

QUARTZ, METAQUARTZITE

SUBARKOSE ►   ◄ SUBLITHARENITE

ARKOSE

LITHIC ARKOSE

FELDSPATHIC LITHARENITE

LITHARENITE

FELDSPARS,
GRANITE + GNEISS
FRAGMENTS

ALL OTHER ROCK
FRAGMENTS
(INCL. CHERT)

FOLK 1968

SANDSTONE SHALE

CALCLITHITE ►   ◄ SANDSTONE,
                        SHALE ARENITES

◄ CHERTARENITE

CARBONATE   CHERT

SEDIMENTARY RF's

VOLCANIC ARENITE ►

◄ SEDARENITE

◄ PHYLLARENITE

VOLCANIC
RF's

METAMORPHIC
RF's

The following photographs illustrate end members of sandstone composition. Several classification names are given for each example.

Devonian Oriskany Ss.
Virginia

Quartzarenite (Folk, 1968); orthoquartzite (Folk, 1954); or quartz arenite (Williams, Turner, and Gilbert, 1954). Some interstitial clay is present but may be, in large part, authigenic. Rock consists mainly of single-crystal, rounded quartz grains.

XN   0.15 mm

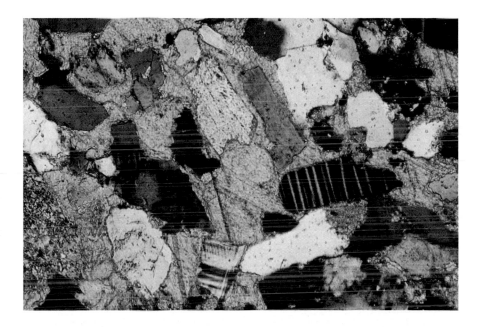

Permian Abo Ss.
New Mexico

A plagioclase arkose (Folk, 1968); arkose (Folk, 1954); or arkosic arenite (Williams, Turner, and Gilbert, 1954). Approximately 25-30 percent of the rock consists of feldspars, with plagioclase somewhat more abundant than microcline. A complete Folk (1968) name for the rock would be— Fine sandstone: calcitic submature micaceous plagioclase arkose.

XN   0.10 mm

Triassic Chinle Fm.
Arizona

A litharenite (shale arenite) (Folk, 1968); subgraywacke (Folk, 1954); or lithic arenite (Williams, Turner, and Gilbert, 1954). Rock has more than 25 percent sedimentary rock fragments with shale clasts slightly more abundant than chert fragments. Interstitial clay composes less than 2 percent of the total rock. Shale clasts were soft and have been strongly deformed.

0.30 mm

Lower Cretaceous Corwin Fm.
Alaska

A chertarenite (Folk, 1968); orthoquartzite (Folk, 1954); or quartz arenite
(Williams, Turner, and Gilbert, 1954).
Rock consists predominantly of detrital
chert fragments which are plotted as quartz
in the Folk (1954) and Williams, Turner,
and Gilbert (1954) classifications but are
counted as sedimentary rock fragments in
the Folk (1968) scheme. In this case, at
least, the Folk (1968) plot yields the
greatest provenance significance.

XN   0.38 mm

Triassic Dockum Fm.
New Mexico

A calclithite (Folk, 1954 and 1968) or
lithic arenite (Williams, Turner, and Gilbert,
1954). The rock consists of over 90 percent limestone fragments (possibly a
reworked caliche) and grains show extensive marginal dissolution and iron oxide
staining. A complete Folk (1968) name
would be—Slightly granular coarse sandstone: hematitic, stylolitized, submature
calclithite.

0.38 mm

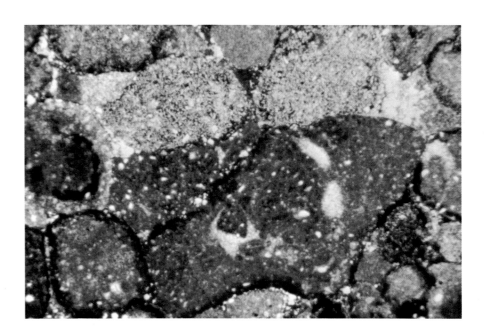

Lower Cretaceous Corwin Fm.
Alaska

A calclithite or chertarenite (Folk, 1968);
orthoquartzite (Folk, 1954); or lithic
arenite (Williams, Turner, and Gilbert,
1954). Subequal chert and carbonate
fragments (both limestone and dolomite)
make up most of this rock. The Folk
(1954) classification ignores carbonate
grains if they compose less than 50 percent
of the sediment—thus, this rock becomes
an orthoquartzite in that scheme.

XN   0.15 mm

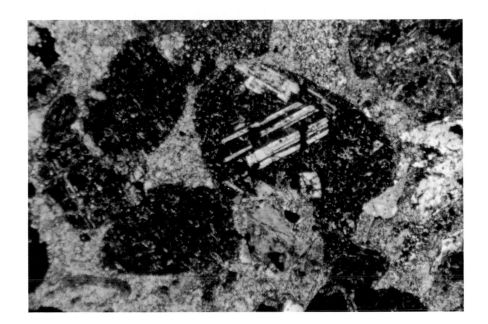

Cretaceous Ildefonso Fm.
Puerto Rico

A volcanic-arenite (Folk, 1968); arkose (Folk, 1954); or lithic arenite (Williams, Turner, and Gilbert, 1954). Rock consists mainly of rounded basic igneous rock fragments. In Folk's 1954 scheme these are grouped with feldspars; in his 1968 classification they are plotted as rock fragments. Because of the remarkable variety of possible names for the same rock using different classifications, reference should always be made to the classification used.

XN   0.30 mm

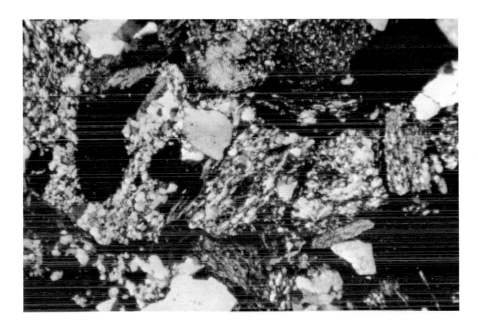

Cretaceous Fortress Mountain Fm.
Alaska

A high-rank phyllarenite (Folk, 1968); graywacke (Folk, 1954); or quartz-lithic arenite (Williams, Turner, and Gilbert, 1954). Rock consists mainly of gneiss, schist, and metaquartzite fragments. The metamorphic source is best expressed in Folk's 1968 classification.

XN   0.38 mm

Cretaceous Tuktu Fm.
Alaska

A low-rank phyllarenite (Folk, 1968); graywacke (Folk, 1954); or lithic arenite (Williams, Turner, and Gilbert, 1954). The rock is composed mainly of phyllite and schist fragments with associated quartz. Chlorite and muscovite are the major micaceous minerals visible in this photo. The sediment has undergone considerable compactional deformation.

XN   0.24 mm

Selected Sandstone Classification Bibliography

Bokman, J., 1952, Clastic quartz particles as indices of provenance: Jour. Sed. Petrology, v. 22, p. 17-24.

Chen, P. -Y., 1968, A modification of sandstone classification: Jour. Sed. Petrology, v. 38, p. 54-60.

Crook, C. A. W., 1960, Classification of arenites: Am. Jour. Sci., v. 258, p. 419-428.

Dapples, E. C., 1947, Sandstone types and their associated depositional environments: Jour. Sed. Petrology, v. 17, p. 91-100.

—— and W. C. Krumbein, L. L. Sloss, 1953, Petrographic and lithologic attributes of sandstones: Jour. Geology, v. 61, p. 291-317.

Dott, R. H., 1964, Wacke, graywacke and matrix—what approach to immature sandstone classification?: Jour. Sed. Petrology, v. 34, p. 625-632.

Fisher, R. V., 1966, Rocks composed of volcanic fragments and their classification: Earth Sci. Rev., v. 1, p. 287-298.

Folk, R. L., 1954, The distinction between grain size and mineral composition in sedimentary rock nomenclature: Jour. Geology, v. 62, p. 344-359.

—— 1956, The role of texture and composition in sandstone classification: Jour. Sed. Petrology, v. 26, p. 166-171.

—— 1968, Petrology of sedimentary rocks: Austin, Texas, Hemphill's Book Store, 170 p.

—— P. B. Andrews, and D. W. Lewis, 1970, Detrital sedimentary rock classification and nomenclature for use in New Zealand: New Zealand Jour. Geol. and Geophys., v. 13, p. 937-968.

Füchtbauer, H., 1959, Zur nomenklatur der sedimentgesteine: Erdöl u. Kohle, v. 12, p. 605-613.

Grabau, A. W., 1904, On the classification of sedimentary rocks: Am. Geologist, v. 33, p. 228-247.

Huckenholtz, H. G., 1963, A contribution to the classification of sandstones: Geol. foren. Stockholm Forh., v. 85, p. 156-172.

Klein, G. DeV., 1963, Analysis and review of sandstone classifications in the North American geological literature, 1940-1960: Geol. Soc. America Bull., v. 74, p. 555-576.

Krumbein, W. C., and L. L. Sloss, 1963, Stratigraphy and sedimentation, 2nd ed.: San Francisco, W. H. Freeman and Co., 660 p.

Krynine, P. D., 1948, The megascopic study and field classification of sedimentary rocks: Jour. Geology, v. 56, p. 130-165.

McBride, E. F., 1963, A classification of common sandstones: Jour. Sed. Petrology, v. 33, p. 664-669.

Okada, H., 1971, Classification of sandstone: analysis and proposal: Jour. Geology, v. 79, p. 509-525.

Packman, G. H., 1954, Sedimentary structures as an important factor in the classification of sandstones: Am. Jour. Sci., v. 252, p. 466-476.

Pettijohn, F. J., 1949, Sedimentary rocks, 1st ed: New York, Harper Bros., 526 p.

—— 1954, Classification of sandstones: Jour. Geology, v. 62, p. 360-365.

—— 1957, Sedimentary rocks, 2nd ed.: New York, Harper Bros., 719 p.

—— P. E. Potter, and R. Siever, 1972, Sand and sandstone: New York, Springer-Verlag, 618 p.

Picard, M. D., 1971, Classification of fine-grained sedimentary rocks: Jour. Sed. Petrology, v. 41, p. 179-195.

Shvetsov, M. S., 1948, Petrografiya osadochnikh porod (Petrography of sedimentary rocks, M. I. Smith, trans.), 2nd ed.: Moscow, Gostoptekhisdat (copy in Univ. Texas Geol. library).

Stone, W. J., and J. M. Erickson, 1970, A FORTRAN program for Folk's sandstone classification: Compass, v. 47, p. 163-168.

Tallman, S. L., 1949, Sandstone types: their abundance and cementing agents: Jour. Geology, v. 57, p. 582-591.

Travis, R. B., 1970, Nomenclature for sedimentary rocks: AAPG Bull., v. 54, p. 1095-1107.

Washburne, A. L., J. E. Sanders, and R. F. Flint, 1963, A convenient nomenclature for poorly sorted sediments: Jour. Sed. Petrology, v. 33, p. 478-480.

Williams, Howel, F. J. Turner, and C. M. Gilbert, 1954, Petrography: San Francisco, W. H. Freeman and Co., 406 p.

Textures

102

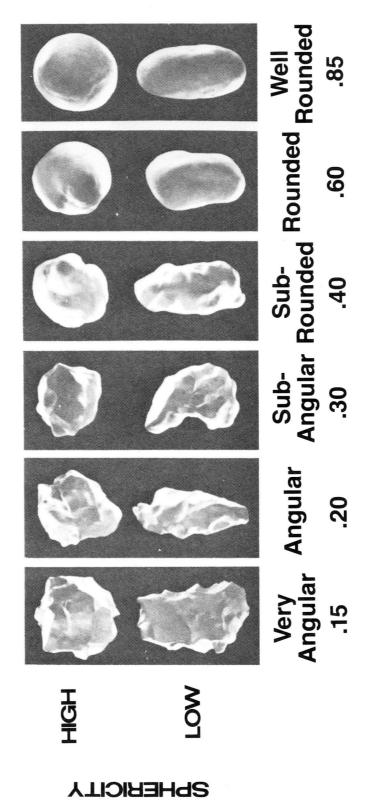

Very Angular .15 | Angular .20 | Sub-Angular .30 | Sub-Rounded .40 | Rounded .60 | Well Rounded .85

HIGH

LOW

SPHERICITY

(from: Powers, 1953)

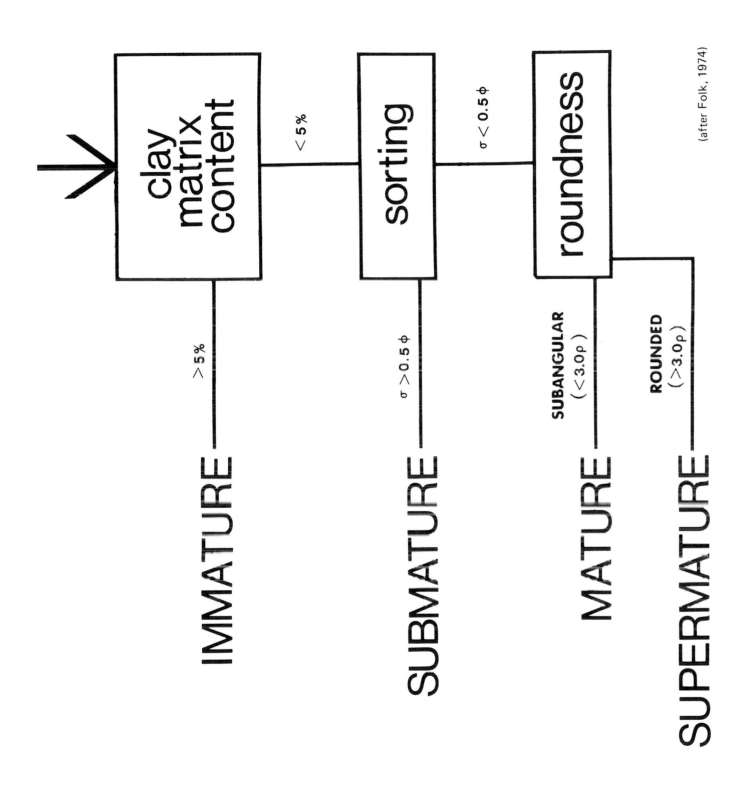

clay matrix content

> 5% — IMMATURE

< 5%

sorting

$\sigma > 0.5\,\phi$ — SUBMATURE

$\sigma < 0.5\,\phi$

roundness

SUBANGULAR ($< 3.0\rho$) — MATURE

ROUNDED ($> 3.0\rho$) — SUPERMATURE

(after Folk, 1974)

Comparison chart for sorting and sorting classes.
(from: Pettijohn, Potter, and Siever, 1972)

SORTING IMAGES

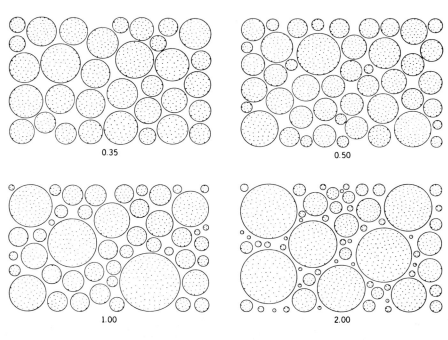

DIAMETER RATIO (MILLIMETERS)	PHI STANDARD DEVIATION	VERBAL SCALE	
1.0	0.00		
		very well sorted	MATURE
1.6	0.35		
		well sorted	
2.0	0.50		
		moderately sorted	
4.0	1.00		SUBMATURE
		poorly sorted	
16.0	2.00		
		very poorly sorted	

(After Folk, 1965, p. 104–105)

Pennsylvanian Teuschnitzer Cgl.
Germany

An immature sedimentary texture (lithic wacke or phyllarenite). Sample contains far more than 5 percent detrital clay matrix, which automatically classes it as an immature sediment. In addition, coarser grains are very poorly sorted and very angular. Sediments this poorly sorted are most commonly encountered in alluvial fan, submarine fan, turbidite, or upper floodplain settings although they can be found to a lesser degree in almost all environments. Considerable care must be taken to use only percentages of detrital, not authigenic, clays in textural-environmental interpretations.

0.28 mm

Permian Abo Ss.
New Mexico

A submature rock (an arkose). This rock has less than 5 percent detrital clay matrix (although more clay is present in detrital rock fragments), and sorting is relatively poor; thus, this sample has a submature texture. The rounding of grains is very poor in this example. Submature textures can indicate relatively low energy depositional environments in almost any sedimentary setting.

0.38 mm

Oligocene Catahoula Tuff *
Texas

A mature rock (an opal-cemented chertarenite). Sample has no clay matrix, is well sorted, and has very poor grain rounding—this defines a mature texture. Mature textures are produced, in most cases, in moderately high energy environments with rapid deposition.

0.24 mm

Jurassic Entrada Ss. ⋆
Utah

A supermature rock (a quartzarenite or orthoquartzite). Sample contains less than 5 percent detrital clay matrix (green epoxy fills pore space), is very well sorted, and is composed of grains which are very well rounded—this defines a supermature texture. Texturally supermature rocks are commonly mineralogically mature as well. Such textures are most frequently produced in high energy settings, especially beaches and eolian dunes.

0.27 mm

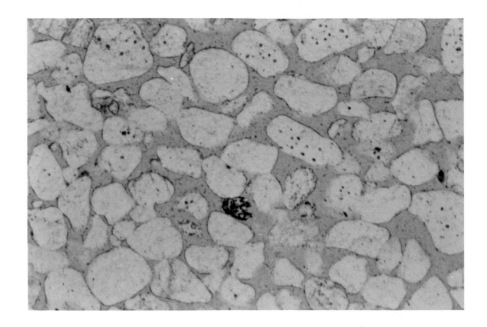

Silurian Clinton Fm.
Virginia

A textural inversion—well rounded but poorly sorted (possibly bimodal) grains. Such textural inversions, while not uncommon, do indicate unusual environmental conditions. Possible causes include mixing of sediment from two different environments, storm mixing of material in a high energy environment, or multiple sources of sand supply.

XN   0.38 mm

Cambrian Hickory Ss. Mbr. of Riley Fm. ⋆
Texas

A textural inversion—a bimodal grain size distribution. Coarse sand grains are mixed together with very fine sand grains with few fragments in the intermediate size ranges. Such a texture can result from the mechanisms outlined in the previous example or, in some cases, can be produced by burrowing or other *in situ* mixing processes.

XN   0.27 mm

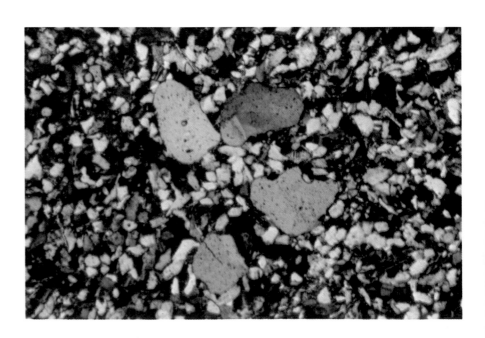

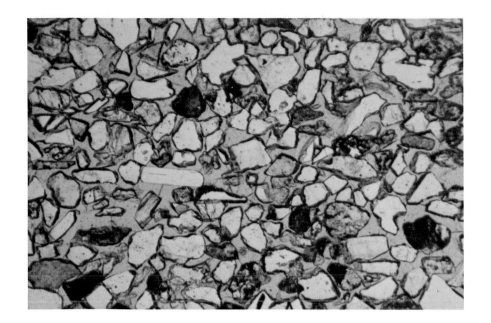

Oligocene Catahoula Tuff
Texas

When oriented thin sections are cut (generally perpendicular to bedding) from oriented rock slabs they can be used to do quantitative or qualitative grain orientation studies. The investigations can be performed directly on the microscope by observing the long axis of grains or by a photometric method which integrates the extinction behavior of all the grains in a field of view.

0.38 mm

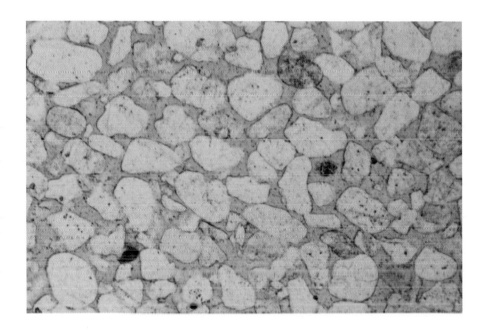

Jurassic Entrada Ss.
Utah

Although orientation studies are most easily performed where grains are elongate and angular, as in the previous example, such studies can also be done on rounded, more equant grains as in this example. It should be remembered that orientation measured in any two-dimensional section is an apparent orientation only. Sections cut perpendicular to each other should be used to establish three-dimensional orientation.

0.27 mm

Upper Cretaceous Mowry Shale
Utah

Grain orientation can be determined in fissile shales as well as in coarser grained sediments. In this example, clay minerals are preferentially oriented and extreme changes are noted in overall birefringence as the stage is rotated with crossed polarizers. Fissility, slaty cleavage, and other similar features can be detected this way.

XN   0.10 mm

Selected Texture Bibliography

Grain Size-Sorting

Allen, Terence, 1968, Particle size measurement: London, Chapman and Hall, 248 p.

Beard, D. C., and P. K. Weyl, 1973, Influence of texture on porosity and permeability of unconsolidated sand: AAPG Bull., v. 57, p. 349-369.

Cadigan, R. A., 1961, Geologic interpretation of grain-size distribution measurements of Colorado Plateau sedimentary rocks: Jour. Geology, v. 69, p. 121-144.

Folk, R. L., 1954, The distinction between grain size and mineral composition in sedimentary rock nomenclature: Jour. Geology, v. 62, p. 344-359.

—— 1966, A review of grain-size parameters: Sedimentology, v. 6, p. 73-93.

—— 1968, Bimodal supermature sandstones: Product of the desert floor, *in* Rept. 23rd Session: Internat. Geol. Cong., Prague, Proc., Sec. 8, p. 9-32.

—— 1974, Petrology of sedimentary rocks: Austin, Texas, Hemphill's Book Store, 170 p.

—— and R. C. Ward, 1957, Brazos River bar: A study in the significance of grain size parameters: Jour. Sed. Petrology, v. 27, p. 3-26.

Friedman, G. M., 1958, Determination of sieve size distribution from thin section data for sedimentary petrological studies: Jour. Geology, v. 66, p. 394-416.

—— 1961, Distinction between dune, beach, and river sands from their textural characteristics: Jour. Sed. Petrology, v. 31, p. 514-529.

—— 1962, Comparison of moment measures for sieving and thin section data in sedimentary petrological studies: Jour. Sed. Petrology, v. 32, p. 15-25.

—— 1967, Dynamic processes and statistical parameters compared for size frequency distribution of beach and river sands: Jour. Sed. Petrology, v. 37, p. 327-354.

Gaither, A., 1953, A study of porosity and grain relationships in experimental sands: Jour. Sed. Petrology, v. 23, p. 180-195.

Inman, D. L., 1949, Sorting of sediment in light of fluid mechanics: Jour. Sed. Petrology, v. 19, p. 51-70.

Klovan, J. E., 1966, The use of factor analysis in determining depositional environments from grain-size distributions: Jour. Sed. Petrology, v. 36, p. 115-125.

—— and J. T. Solohub, 1968, Grain-size parameters: A critical evaluation of their significance: Geol. Soc. America Annual Meeting Abstracts, Mexico City, p. 161-162.

Koldijk, W. S., 1968, On environment-sensitive grain-size parameters: Sedimentology, v. 10, p. 57-69.

Mason, C. C., and R. L. Folk, 1958, Differentiation of beach, dune and aeolian flat environments by size analysis, Mustang Island, Texas: Jour. Sed. Petrology, v. 28, p. 211-226.

Middleton, G. V., 1962, Size and sphericity of quartz grains in two turbidite formations: Jour. Sed. Petrology, v. 32, p. 725-742.

Moiola, R. J., and D. Weiser, 1968, Textural parameters: An evaluation: Jour. Sed. Petrology, v. 38, p. 45-53.

Morrow, N. R., J. D. Huppler, and A. B. Simmons, III, 1969, Porosity and permeability of unconsolidated, Upper Miocene sands from grain-size analysis: Jour. Sed. Petrology, v. 39, p. 312-321.

Passega, Renato, 1957, Texture as characteristic of clastic deposition: AAPG Bull., v. 41, p. 1952-1984.

—— 1964, Grain size representation by CM patterns as a geological tool: Jour. Sed. Petrology, v. 34, p. 830-847.

Potter, J. F., 1977, The texture of compacted bimodal sediments: Jour. Sed. Petrology, v. 47, p. 1539-1541.

Powers, M. C., 1953, A new roundness scale for sedimentary particles: Jour. Sed. Petrology, v. 23, p. 117-119.

Pryor, W. A., 1973, Permeability-porosity patterns and variations in some Holocene sand bodies: AAPG Bull., v. 57, no. 1, p. 162-189.

Reed, W. E., R. L. Fever, and G. J. Moir, 1975, Depositional environment interpretation from settling-velocity (Psi) distributions: Geol. Soc. America Bull., v. 86, p. 1321-1328.

Rosenfeld, M. A., Lynn Jacobsen, and J. C. Ferm, 1953, A comparison of sieve and thin-section technique for size analysis: Jour. Geology, v. 61, p. 114-132.

Sevon, W. D., 1966, Distinction of New Zealand beach, dune and river sands by their grain size distribution characteristics: New Zealand Jour. Geol. and Geophys., v. 9, p. 212-223.

Shepard, F. P., and R. Young, 1961, Distinguishing between beach and dune sands: Jour. Sed. Petrology, v. 31, p. 196-214.

Swift, D. J. P., J. R. Schubel, and R. W. Sheldon, 1972, Size analysis of fine-grained suspended sediments: a review: Jour. Sed. Petrology, v. 42, p. 122-134.

Tillman, R. W., 1971, Multiple group discriminant analysis of grain size data as an aid in recognizing environments of deposition: VIIIth Internat. Sediment. Cong., Heidelberg, Abstracts volume, p. 102.

Visher, G. S., 1965, Fluvial processes as interpreted from ancient and recent fluvial deposits, *in* G. V. Middleton, ed., Primary sedimentary structures and their hydrodynamic interpretation: Soc. Econ. Paleontologists and Mineralogists Spec. Pub. no. 12, p. 116-132.

—— 1969, Grain size distributions and depositional processes: Jour. Sed. Petrology, v. 39, p. 1074-1106.

Wentworth, C. K., 1922, A scale of grade and class terms for clastic sediments: Jour. Geology, v. 30, p. 377-392.

Wolff, R. G., 1964, The dearth of certain sizes of materials in sediments: Jour. Sed. Petrology, v. 34, p. 320-327.

Grain Shape—Abrasion

Anderson, G. E., 1926, Experiments on the rate of wear of sand grains: Jour. Geology, v. 34, p. 144-158.

Blatt, H., 1959, Effect of size and genetic quartz type on sphericity and form of beach sediments, northern New Jersey: Jour. Sed. Petrology, v. 29, p. 197-206.

Bradley, W. C., 1970, Effect of weathering on abrasion of granitic gravel, Colorado River, Texas: Geol. Soc. America Bull., v. 81, p. 61-80.

Briggs, L. I., D. S. McCulloch, and F. Moser, 1962, The hydraulic shape of sand particles: Jour. Sed. Petrology, v. 32, p. 645-657.

—— F. Moser, and D. S. McCulloch, 1961, Hydraulic shape of mineral grains (abs.): Geol. Soc. America Spec. Paper 68, p. 140.

Crook, K. A. W., 1968, Weathering and roundness of quartz sand grains: Sedimentology, v. 11, p. 171-182.

Feniak, M. W., 1944, Grain sizes and shapes of various minerals in igneous rocks: Am. Mineralogist, v. 29, p. 415-421.

Folk, R. L., 1951, Stages of textural maturity in sedimentary rocks: Jour. Sed. Petrology, v. 21, p. 127-130.

—— 1955, Student operator error in determination of roundness, sphericity, and grain size: Jour. Sed. Petrology, v. 25, p. 297-301.

Galloway, J. J., 1919, The rounding of sand grains by solution: Am. Jour. Sci., v. 47, p. 270-280.

Knight, S. H., 1924, Eolian abrasion of quartz grains (abs.): Geol. Soc. America Bull., v. 35, p. 107.

Krumbein, W. C., 1941, The effects of abrasion on the size, shape, and roundness of rock fragments: Jour. Geology, v. 49, p. 482-520.

Kuenen, P. H., 1958, Some experiments on fluviatile rounding: Kon. Ned. Akad. Wetensch. Amsterdam, Proc., series B., v. 61, p. 47-53.

—— 1959, Experimental abrasion: 3. Fluviatile action on sand: Am. Jour. Sci., v. 257, p. 172-190.

MacCarthy, G. R., 1933, The rounding of beach sands: Am. Jour. Sci., v. 25, p. 205-224.

—— and J. W. Hubble, 1938, Shape sorting of sand grains by wind action: Am. Jour. Sci., v. 35, p. 64-73.

Mansland, P. S., and J. G. Woodruff, 1937, A study of the effects of wind transportation on grains of several minerals: Jour. Sed. Petrology, v. 7, p. 18-30.

McIntyre, D. D., 1959, The hydraulic equivalence and size distributions of some mineral grains from a beach: Jour. Geology, v. 67, p. 278-301.

Moss, A. J., 1972, Initial fluviatile fragmentation of granitic quartz: Jour. Sed. Petrology, v. 42, p. 905-916.

Pearce, T. H., 1971, Short distance fluvial rounding of volcanic detritus: Jour. Sed. Petrology, v. 41, p. 1069-1072.

Powers, M. C., 1953, A new roundness scale for sedimentary particles: Jour. Sed. Petrology, v. 23, p. 117-119.

Russell, R. D., 1939, Effects of transportation of sedimentary particles, in P. D. Trask, ed., Recent marine sediments: Tulsa, AAPG, p. 32-47.

—— and R. E. Taylor, 1937, Roundness and shape of Mississippi River sands: Jour. Geology, v. 45, p. 225-267.

Shukis, P. S., and F. G. Ethridge, 1975, A petrographic reconnaissance of sand size sediment, upper St. Francis River, southeastern Missouri: Jour. Sed. Petrology, v. 45, p. 115-127.

Smith, N. D., 1972, Flume experiments on the durability of mud clasts: Jour. Sed. Petrology, v. 42, p. 378-383.

Sneed, E. D., and R. L. Folk, 1958, Pebbles in the lower Colorado River, Texas; a study in particle morphogenesis: Jour. Geology, v. 66, p. 114-150.

Swan, Bernard, 1974, Measures of particle roundness: a note: Jour. Sed. Petrology, v. 44, p. 572-577.

Thiel, G. A., 1940, The relative resistance to abrasion of mineral grains of sand size: Jour. Sed. Petrology, v. 10, p. 103-124.

Waskom, J. D., 1958, Roundness as an indicator of environment along the coast of panhandle Florida: Jour. Sed. Petrology, v. 28, p. 351-360.

Grain Orientation

Bonham, L. C., and J. H. Spotts, 1971, Measurement of grain orientation, in R. E. Carver, ed., Procedures in sedimentary petrology: New York, Wiley-Interscience, p. 285-312.

Colburn, I. P., 1968, Grain fabrics in turbidite sandstone beds and their relationship to sole mark trends on the same bed: Jour. Sed. Petrology, v. 38, p. 146-158.

Curray, J. R., 1956, Dimensional grain orientation studies of recent coastal sands: AAPG Bull., v. 40, p. 2440-2456.

Gibbons, G. S., 1972, Sandstone imbrication study in planar sections: Dispersion, biases, and measuring methods: Jour. Sed. Petrology, v. 42, p. 966-972.

Kahn, J. S., 1956, The analysis and distribution of packing in sand-size sediments: Jour. Geology, v. 64, p. 385-395.

Krumbein, W. C., 1939, Preferred orientation of pebbles in sedimentary deposits: Jour. Geology, v. 47, p. 673-706.

Martini, I. P., 1971, A test of validity of quartz grain orientation as a paleocurrent and paleoenvironment indicator: Jour. Sed. Petrology, v. 41, p. 60-68.

Onions, D., and G. V. Middleton, 1968, Dimensional grain orientation of Ordovician turbidite graywackes: Jour. Sed. Petrology, v. 38, p. 164-174.

Parkash, B., and G. V. Middleton, 1969, Grain and graptolite orientation in turbidite graywacke, Cloridorme Formation (Ordovician), Gaspe, Quebec: AAPG Bull., v. 53, p. 735.

Rad, U. von, 1970, Comparison between "magnetic" and sedimentary fabric in graded and cross laminated sand layers, southern California: Geol. Rundschau, v. 60, p. 331-354.

Sedimentary Petrology Seminar, 1965, Gravel fabric in Wolf Run: Sedimentology, v. 4, p. 273-203.

Sestini, G., and G. Pranzini, 1965, Correlation of sedimentary fabric and sole marks as current indicators in turbidites: Jour. Sed. Petrology, v. 35, p. 100-108.

Shelton, J. W., and D. E. Mack, 1970, Grain orientation in determination of paleocurrents and sandstone trends: AAPG Bull., v. 54, p. 1108-1119.

Sippel, R. F., 1971, Quartz grain orientations—1 (the photometric method): Jour. Sed. Petrology, v. 41, p. 38-59.

Spotts, J. H., 1964, Grain orientation and imbrication in Miocene turbidity current sandstones, California: Jour. Sed. Petrology, v. 34, p. 229-253.

Cements

Pennsylvanian Ricker Ss. Mbr. of Mineral
  Wells Fm.
Texas

An early to intermediate stage of quartz
overgrowth cementation. The detrital
quartz grains have well developed, clearly
visible, idiomorphic overgrowths of quartz
which formed in optical continuity with
the underlying quartz. These overgrowths
are readily visible because of abundant
inclusions trapped at the boundary between detrital and authigenic quartz and
because of the euhedral terminations of
the overgrowth.

XN   0.08 mm

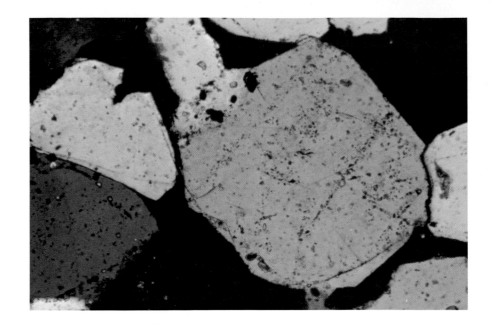

Ordovician St. Peter Ss. ★
Arkansas

Early stages of quartz overgrowth cementation seen in SEM. Numerous very small
overgrowths give appearance of pitted
surface on detrital quartz grains. Larger
overgrowths with smooth crystal faces can
be seen to be forming simultaneously at
different places on the grain surface. All
overgrowths are, however, oriented in the
same crystallographic direction as the
detrital nucleus. Photo by E. D. Pittman.

SEM   55 μm

Pennsylvanian "Gray Ss." (Strawn Gp.)
  1,398 m (4,586 ft)
Texas

Intermediate to advanced stage of quartz
overgrowth cementation in which a significant amount of pore space (shown in blue)
has been filled. Note euhedral crystal faces
where overgrowths do not interfere with
each other, partial visibility of inclusions
or "dust rims" beneath overgrowths, and
additional calcite cement. Photo by S. P.
Dutton.

0.09 mm

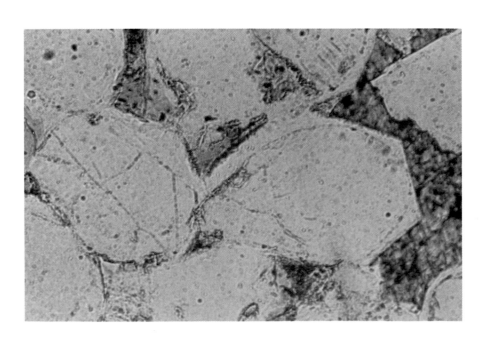

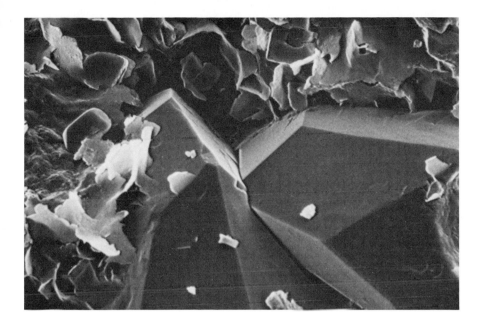

Upper Permian Bell Canyon Fm.
Texas 1,396 m (4,580 ft)

Intermediate to advanced stage of quartz overgrowth cementation seen in SEM. Individual small overgrowths have coalesced into single, continuous, euhedral overgrowths which encompass the entire quartz grain and fill a significant portion of the total porosity. Quartz cementation is partly contemporaneous with, and partly prior to, clay cementation. Photo by C. R. Williamson.

SEM   10 μm

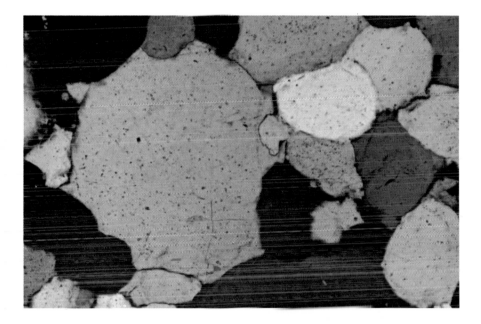

Cambrian Gatesburg Fm. ⋆
Pennsylvania

Advanced stage of overgrowth cementation in which almost all pore space has been filled by quartz cement. Boundaries between detrital grain nuclei and overgrowth cement are only weakly visible. Contacts between adjacent overgrowths are irregular compromise boundaries produced by mutual interference during crystal growth.

XN   0.15 mm

Ordovician Calico Rock Ss. Mbr.
 of Everton Fm. ⋆
Arkansas

Advanced stage of overgrowth cementation seen in SEM. Interlocking crystal boundaries, lack of porosity, and numerous crystal faces which virtually completely mask original detrital grains are characteristic of this stage of diagenetic alteration. Photo by E. D. Pittman.

SEM   100 μm

Devonian Hoing Ss. Mbr. of Cedar
 Valley Ls. *
Illinois

Quartz overgrowths are commonly difficult
to recognize in thin section, particularly
where "dust rims" and crystal terminations
are scarce or absent. In this example,
sutured boundaries between grains might
well be interpreted as due to compaction
rather than cementation, although careful
examination shows thin "dust rims" and
euhedral crystal terminations to be present.
Photo by R. F. Sippel.

0.10 mm

Devonian Hoing Ss. Mbr. of Cedar
 Valley Ls. *
Illinois

Same as previous photo but with cathodoluminescence. Detrital quartz grains luminesce (orange and blue) and are well
rounded. Angular, euhedral quartz overgrowths do not luminesce (because of
different trace element composition) and
thus contrast sharply with the detrital
grains. This technique often can be very
useful in distinguishing between cementation and compaction fabrics. Photo by R.
F. Sippel.

CL   0.10 mm

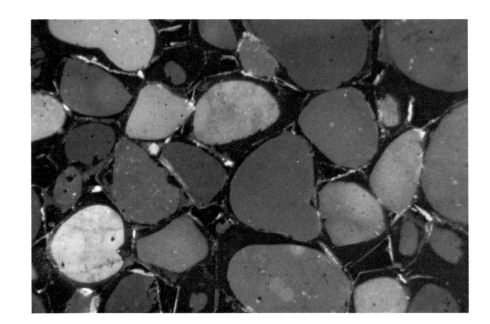

Jurassic Curtis Fm.
Utah

Quartz overgrowth cements are, in some
instances, selectively located on grains. In
this example, quartz cementation is concentrated near grain contacts in a
"meniscus" fabric. This texture has been
interpreted to be the result of near-surface
cementation in the vadose zone. In that
setting, water droplets (and thus cements,
as well) are concentrated and held at grain
contacts.

XN   0.08 mm

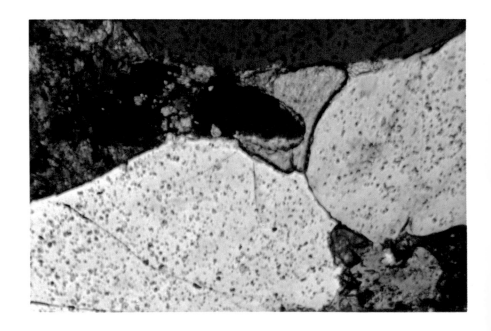

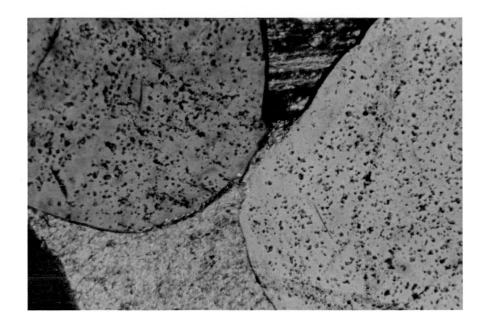

Jurassic Curtis Fm.
Utah

Detail of meniscus quartz cement. Overgrowth quartz is strongly concentrated near grain contacts, has curved exterior surfaces, and is in optical continuity with the underlying detrital quartz. Remnant pore space was filled with later sparry calcite cement.

XN   0.06 mm

Cretaceous Travis Peak Fm.
Texas

Quartz grains with overgrowths can be reworked into younger sediments. In this example, the detrital quartz grain in center has overgrowths which were formed before deposition of this sediment (during a previous sedimentary cycle). An important criterion for recognition of multicycle sediment, detrital overgrowths are identified by the absence of interlocking of any overgrowths, the presence of overgrowths only on isolated grains, and (rarely) by the presence of rounded or broken terminations

XN   0.10 mm

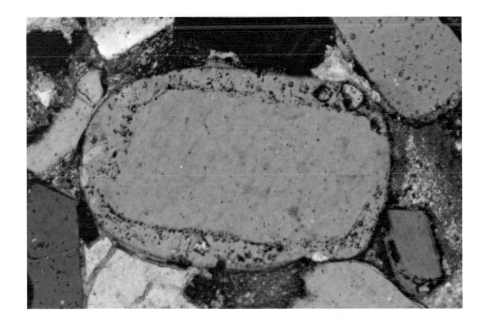

Tertiary 'Vieja Group'
Texas

Quartz grains derived from volcanic source areas may show zonation and rounding which closely mimics quartz overgrowth cement or second-cycle overgrown grains. In this example, the abundant microlites and vacuoles in the outer rim allow recognition of a volcanic source. Quartz overgrowth cements are normally inclusion poor.

XN   0.08 mm

116

Oligocene Catahoula Tuff *
Texas

Opal cemented sandstone. Note the brownish color of cementing opal and the very high relief of all the detrital grains enclosed in the cement. Both of these effects are characteristic of opal cement and are produced by the abundance of waterfilled inclusions in the opal. The inclusions account for the very low refractive index of opal which is the most easily used identifying trait.

0.38 mm

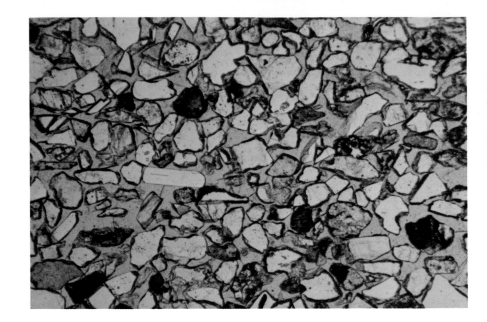

Oligocene Catahoula Tuff *
Texas

Same view as previous photo but with crossed polarizers. Opal cement is completely isotropic and surrounds quartz, chert, feldspar, and other detrital grains, some of which are certainly of volcanic origin. Opal cement is frequently associated with volcaniclastic (siliceous) sediments. Opal cements are chemically metastable and will, in time, dissolve or convert to more stable quartz cement.

XN   0.38 mm

Upper Cambrian Mines Dolomite Mbr.
 of Gatesburg Fm.
Pennsylvania

Complex quartz cementation in a silicified oolitic limestone. Chertified ooids are surrounded by a fringe of fibrous-to-bladed chalcedony and the remainder of the original pore space is filled with equant megaquartz. Crystal size of cements increases toward the center of former voids, a common feature in cavity-filling fabrics.

XN   0.38 mm

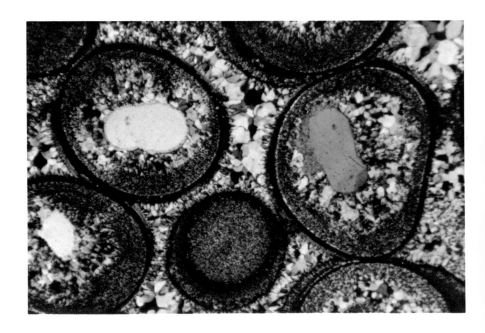

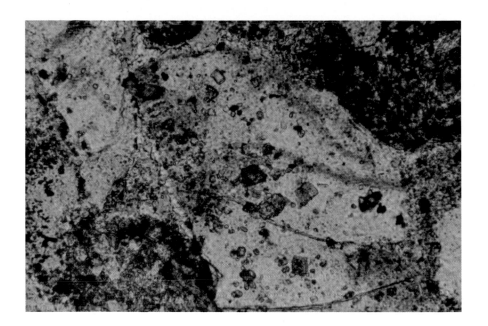

Cretaceous Corwin Fm.
Alaska

The detrital chert grain in the center of photo has typical brownish color of chert and contains numerous dolomite inclusions. Around the margins of the chert grain at left and top, one can see a lighter colored, jagged line of quartz overgrowth crystals. Because the chert is polycrystalline, the overgrowth is polycrystalline as well. Microquartz crystals along the periphery of the chert probably acted as nuclei for quartz cement growth. Porosity filled with green epoxy.

0.06 mm

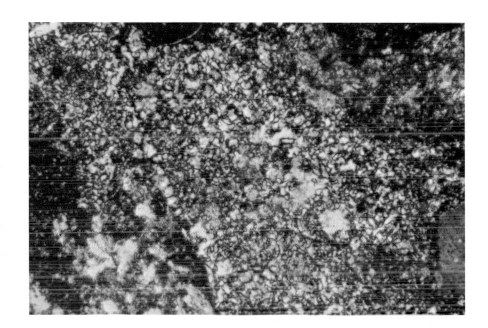

Cretaceous Corwin Fm.
Alaska

Same as previous photo but with crossed polarizers. Rhombic inclusions within chert have very high birefringence and are dolomite. Polycrystalline nature of both the chert and the cement are apparent. Chert is a relatively common cementing agent although chert overgrowths are infrequently seen.

XN   0.06 mm

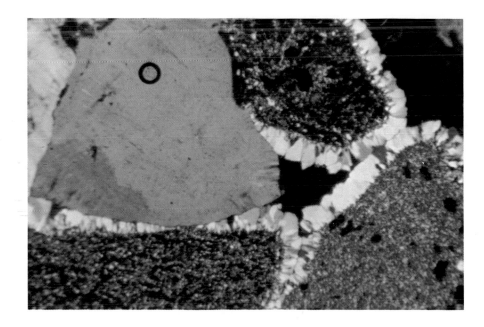

Cretaceous Travis Peak Fm.
Texas

Chert and quartz grains cemented by megaquartz. Each detrital grain has an oriented rim of bladed to equant megaquartz (drusy quartz). Megaquartz (as opposed to microquartz) is defined as having crystals larger than 20 μm. Megaquartz generally forms as a void-filling fabric but it can also be a replacement.

XN   0.38 mm

Eocene Green River Fm.
Wyoming

A chalcedony cemented oolitic carbonate sediment. Chalcedonic quartz consists of small fans of extremely thin, radially oriented quartz fibers. It occurs most commonly, as in this example, as a void filling but is also, more rarely, found as a replacement. This type of chalcedony, which shows variations in extinction along the fiber direction, has been termed "zebraic chalcedony" and has been linked to evaporite formation.

XN   0.38 mm

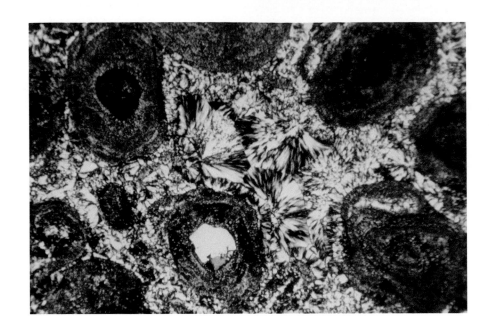

Upper Cretaceous Kogosukruk Tongue
 of Prince Creek Fm.
Alaska

Feldspar overgrowths. A partly altered feldspar has been surrounded by an authigenic overgrowth. The overgrowth is largely inclusion free and unaltered and is definitely authigenic because of its interlocking relationships with adjacent grains. Feldspar overgrowths, although not uncommon, rarely are a major factor in sandstone cementation. Most authigenic feldspar is a pure Na- or K-spar end member.

XN   0.06 mm

Upper Cretaceous Kogosukruk Tongue
 of Prince Creek Fm. *
Alaska

Feldspar overgrowths on several grains. The partially altered detrital cores can be readily distinguished by their abundant liquid-filled vacuoles. These cores are surrounded by clear, unaltered overgrowths which have formed in optical continuity with the detrital grains. Overgrowths interlock slightly with the surrounding grains and thus are demonstrably authigenic. Feldspar overgrowths can be mistaken for quartz overgrowths in some cases.

XN   0.15 mm

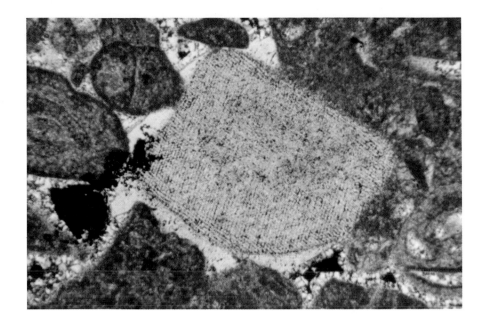

Eocene Ocala Ls.
Florida

Carbonate grains can also form overgrowth cements. In this example, an echinoderm fragment (recognizable by the regular pattern of internal pores and the single-crystal extinction) has a large calcite overgrowth in optical continuity with the core. The overgrowth is clear (inclusion poor) and has irregular outlines with some visible crystal faces. Such overgrowths commonly form early and destroy considerable porosity. Note later silica cement lining pore spaces.

XN   0.30 mm

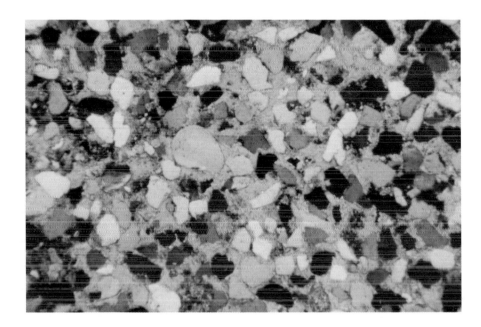

Pennsylvanian Avis Ss. Mbr. of Graham Fm.
Texas

A calcite cemented sandstone. All detrital sand grains in this example are held by a single calcite crystal (note uniformity of calcite extinction) which has completely obliterated porosity. This texture is termed "poikilotopic" cementation and the rock is often referred to as "luster mottled". Commonly the rock will have areas tightly cemented by poikilotopic calcite and other areas lacking all cement—this may be related to the occurrence and distribution of original shell debris in the sediment.

XN   0.30 mm

Permian Abo Ss.
New Mexico

A calcite cemented sandstone in which small patches of poikilotopic cement are present but in which most of the grains are cemented by individual smaller crystals of calcite. In general, the size of the calcite rhombs is closely related to the size of the detrital grains (and the resulting pore spaces).

XN   0.38 mm

Permian Abo Ss.
New Mexico

Detail of a portion of a small patch of poikilotopic calcite cement in a sandstone. Note the uniformity of calcite extinction and the consistency of orientation of twin lamellae throughout the cemented area—both indicate that the cement is a single crystal. Patchy poikilotopic cement is commonly related to the presence of irregularly distributed nuclei for calcite crystallization (shell fragments, pellets, etc.).

XN   0.08 mm

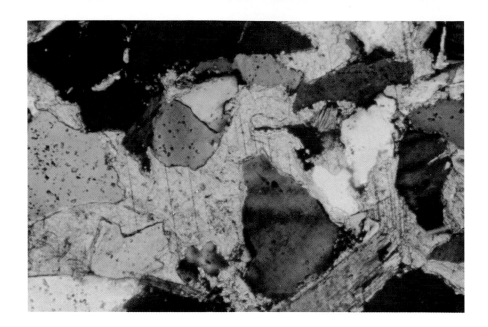

Upper Cretaceous Star Point Ss.
  (Panther Tongue)
Utah

Calcite cementation post-dating quartz overgrowths. Calcite is stained red in this slide to help distinguish it from dolomite. Porosity is shown in blue. Calcite cementation can be a rather late diagenetic occurrence which may destroy a significant amount of porosity. However, calcite cement is also susceptible to later diagenetic dissolution and the formation of secondary porosity.

0.06 mm

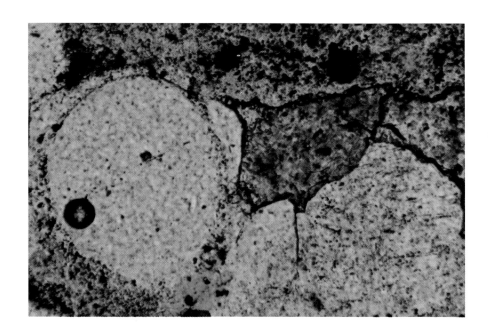

Holocene submarine sands
Persian Gulf

Carbonate cements can have textures which may provide information about their time and place of origin. This example shows radial, fibrous crusts of aragonite cement which have nucleated on thin oolitic coatings around detrital quartz grains. This fabric is characteristic of submarine to intertidal cements produced in marine settings contemporaneous with deposition of the sediment.

0.10 mm

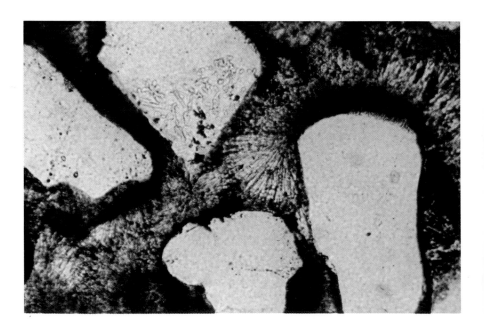

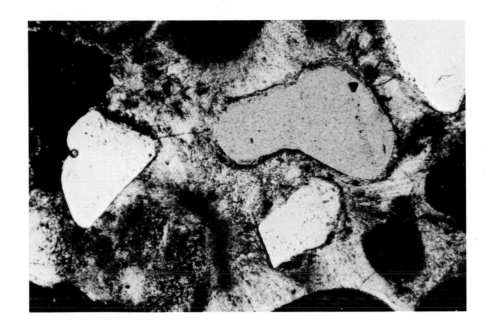

Holocene submarine sands
Persian Gulf

Another example of submarine cementation of sandstone by aragonite. Fibrous aragonite crystals are densely packed in radiating clusters and have nucleated preferentially on carbonate rather than quartz substrates. Although this is a very characteristic and recognizable fabric, aragonite is unstable in most geologic settings and this cement will alter to calcite or will be removed by dissolution with time. The radial fibrous fabric may be preserved in some cases, however.

XN   0.22 mm

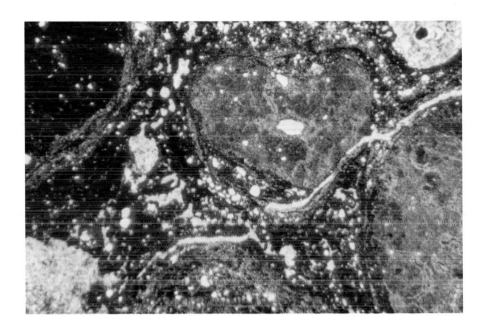

Quaternary caliche crust
Texas

An irregularly laminated, pisolitic, calcitic crust binds carbonate and noncarbonate grains together in this caliche (calcrete). Such crusts, which form in arid climates in most cases, are important indicators of subaerial exposure and weathering. The thin, brownish, laminated crusts, fracturing of grains, marginal dissolution of quartz, pisolites, and interlocking grain textures are aids in recognition of caliche crusts.

1.24 mm

Lower Triassic Bunter Ss.
Northern Ireland

Dolomite cement in a hematitic sandstone. Dolomite is generally characterized by euhedral, rhombic crystal outlines; lack of twinning, and very high birefringence. It is difficult to distinguish from calcite which has very similar birefringence, crystallographic structure, and optical properties. Although the euhedral shape and lack of twinning of dolomite help to distinguish it from calcite, staining is the most accurate and rapid identification tool.

XN   0.10 mm

Upper Mississippian-Permian Nuka Fm.
Alaska

A dolomite cemented sandstone. Dolomite is recognizable in this case by its rhombic outline, lack of twinning, and sharply defined zonation. The zoning consists of alternating intervals of iron-rich (ankerite) and iron-poor dolomite. The cores of these dolomite crystals may be detrital, but the outer zones are authigenic as evidenced by the fact that they partly interlock and surround detrital grains.

XN   0.24 mm

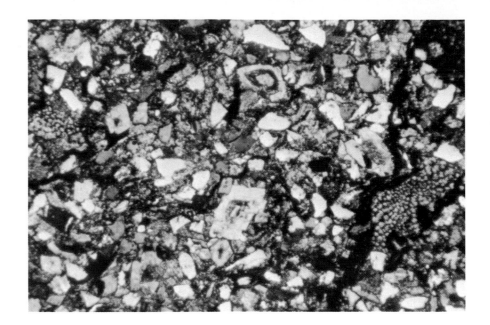

Upper Permian Bell Canyon Fm.
Texas 1,507 m (4,943 ft)

Pore filling authigenic dolomite and clay in a sandstone. In SEM, dolomite commonly shows a euhedral, rhombic outline. In this example, the dolomite growth was subsequent to the formation of chloritic clay cements. Photo by C. R. Williamson.

SEM   8 μm

Pennsylvanian Tensleep Ss.
Wyoming 1,967 m (6,455 ft)

Calcite- and dolomite-cemented sandstone. The calcite cement has been stained red, the dolomite cement is distinguished by moderate relief, rhombic outline, brownish color and lack of response to stain. Note patchy distribution of calcite cement, a common feature in many sandstones. Blue stain marks pore space.

0.27 mm

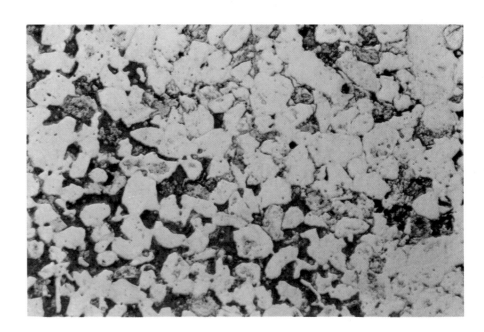

Pennsylvanian Tensleep Ss. ★
Wyoming 1,967 m (6,455 ft)

Detail of previous example showing association of dolomite and calcite cements in
sandstone pore space. Staining is an inex-
pensive, convenient, and invaluable tool for
the distinction of these two petrographically similar minerals. Calcite is stained red by
Alizarin Red whereas dolomite remains
unaffected. Without staining the distinction of these two minerals would be very
difficult in this example.

XN   0.10 mm

Lower Cretaceous sandstone
COST No. B-2 well
U.S. Atlantic offshore 3,911 m (12,830 ft)

Siderite cemented sandstone. Siderite is a
common cement in subsurface sandstone
sections although it is frequently misidentified or overlooked. The finely crystalline
siderite in this example shows a characteristic "flattened rhomb" outline. Siderite
has both indices of refraction well above
that of balsam and thus does not "twinkle"
with stage rotation.

0.06 mm

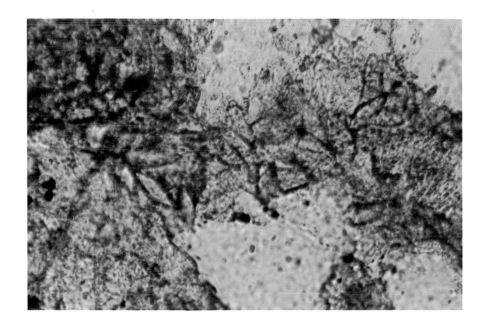

Lower Cretaceous sandstone
COST No. B-2 well
U.S. Atlantic offshore 3,911 m (12,830 ft)

Detail of previous example illustrating the
flattened rhombic outline and typical
brownish color of siderite. Siderite, like
both calcite and dolomite, has a uniaxial
negative figure and extreme birefringence.
Yellow brown color is a result of partial
alteration to limonite. Siderite is an indicator of moderate to strongly reducing
conditions during formation.

0.05 mm

Upper Cretaceous Upper Logan
  Canyon Fm.
Canada (Scotian Shelf) 1,683 m (5,520 ft)

Early diagenetic siderite cement in a quartzarenitic sandstone. Small siderite crystals line pore spaces and perhaps marginally replace quartz grains. Remnant porosity is shown in blue. Crystal outlines, relatively high relief, and brownish color can be used to identify siderite in this example. Photo by D. A. McDonald.

0.40 mm

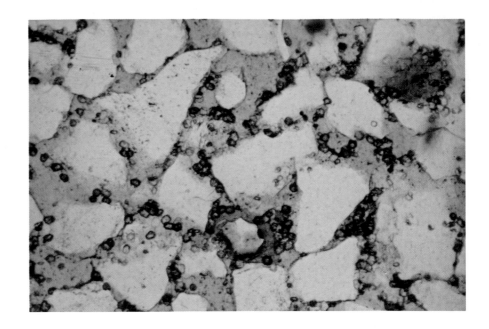

Upper Cretaceous Woodbine Fm. ★
Texas

Hematite-cemented sandstone. This is a section thinner than the standard 30 μm. Normally hematite would be visible only as an opaque filling between grains. In very thin sections and with intense transmitted light, however, one can sometimes see, as here, bladed crusts of deep red hematite crystals. Hematite is generally indicative of oxidizing conditions although it can form at high pH in mildly reducing environments.

XN   0.38 mm

Upper Triassic New Haven Arkose
Connecticut

A more common form of hematite cementation than that shown in previous example is illustrated here. Thin crusts, flakes, or scales of hematite, commonly interspersed with clays, line quartz and other detrital grains. Hematite is apparently mainly derived from the breakdown of unstable iron-bearing heavy minerals and even small amounts of such cement can yield a strongly pigmented redbed sediment.

0.06 mm

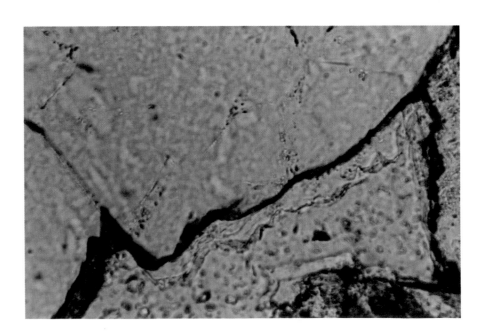

Miocene 'Hayner Ranch Fm.'
New Mexico

Extensive hematite cement in association with a partially corroded detrital hornblende crystal. Intrastratal dissolution of unstable, detrital ferromagnesian minerals is an important source of diagenetic iron oxides, especially in sediments deposited in arid conditions where such unstable detrital grains were not removed during initial weathering. Porosity and translucent grains both appear pale blue in this photo. Photo by T. R. Walker.

RTL   0.07 mm

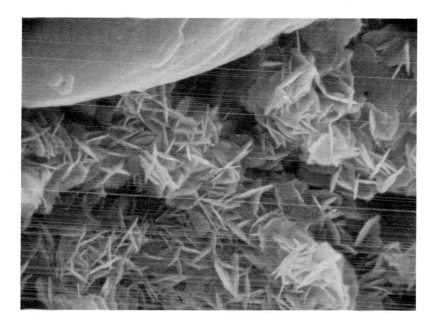

Triassic Moenkopi Fm.
Arizona

SEM view of finely crystalline, red, authigenic hematite coatings on detrital sand grains. The hematite occurs as very finely crystalline platy material which is aggregated into rosette-like clusters not unlike some clay minerals, especially chlorite. Photo by T. R. Walker.

SEM   1.0 μm

Triassic Moenkopi Fm.
Arizona

Finely crystalline and coarsely crystalline authigenic hematite as seen in SEM. The coarser material occurs as platy, hexagonal crystals (similar in appearance to kaolinite crystals), and is dark red to black in reflected light. The more finely crystalline hematite is similar to that shown in the previous photo and appears medium red in reflected light. Photo by T. R. Walker.

SEM   1.2 μm

Cambrian Hickory Ss. Mbr. of Riley Fm.
Texas

Hematite cements can form quite late in
the diagenetic history of sediments. In this
example, thin hematitic films coat numerous hairline fractures produced by intense
deformation (almost mylonitization) of the
sediment. Thus, hematite formation can be
dated as clearly post-deformational.

0.22 mm

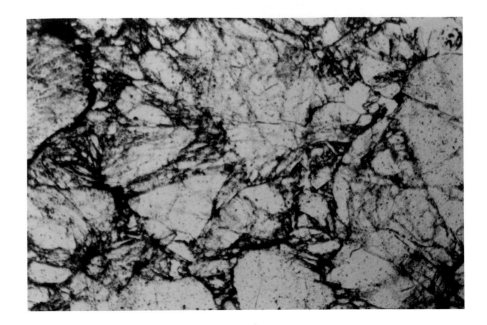

Pleistocene Colorado River terrace deposits
Texas

A limonite- (goethite-) cemented sandstone.
Rounded quartz and microcline grains are
coated by crusts of pore-filling limonite.
In intense transmitted light (especially
with the conoscopic condenser in place)
limonite appears yellow brown. Limonite
is commonly associated with continental
sediments, weathering profiles, and the
oxidation of precursor iron-rich minerals.

XN   0.38 mm

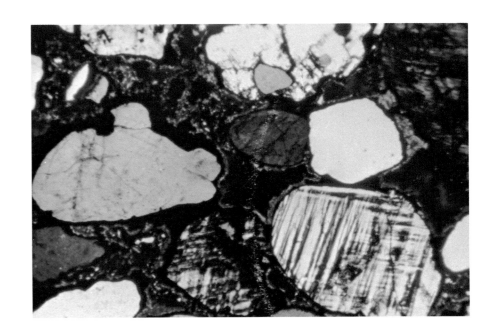

Pliocene Bone Valley Fm.
Florida

A phosphate-cemented sandstone containing quartz and phosphate clasts. Detrital
phosphate grains, in this example, appear
completely isotropic or have extremely low
birefringence. Phosphatic cement, composed primarily of wavellite, has moderate
birefringence and radiating, fibrous crystal
structure. Phosphate cements are most
common in sediments with abundant
detrital phosphate or in association with
hiatus intervals.

XN   0.10 mm

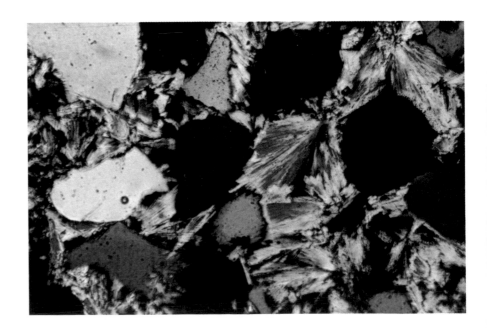

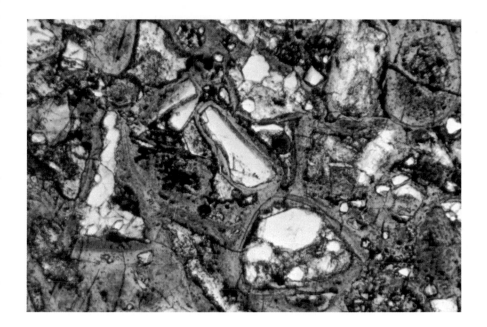

Permian Phosphoria Fm.
Idaho

A phosphate-cemented sandstone. The rock consists of quartz, bone fragments, and other grains oolitically coated by phosphate and then completely cemented by further phosphate and carbonate precipitation. The brownish color is typical of phosphate (collophane or other apatite minerals). Phosphate deposits of this type may mark zones of upwelling of cool, phosphate-saturated, deep-ocean waters on ancient continental margins, other zones of high biological productivity, or hiatus intervals.

0.24 mm

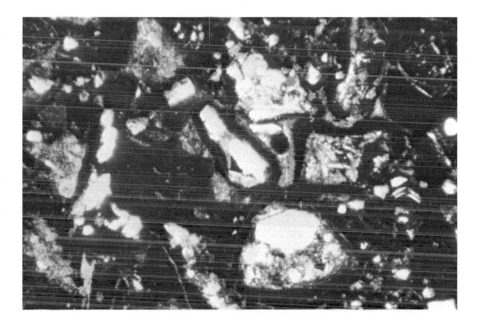

Permian Phosphoria Fm.
Idaho

Same as previous photo but with crossed polarizers. Phosphatic oolitic coatings and phosphatic cement can be seen to be virtually isotropic; phosphatic bone fragments found as nuclei have very low birefringence. The birefringent material filling residual porosity is calcite.

XN   0.24 mm

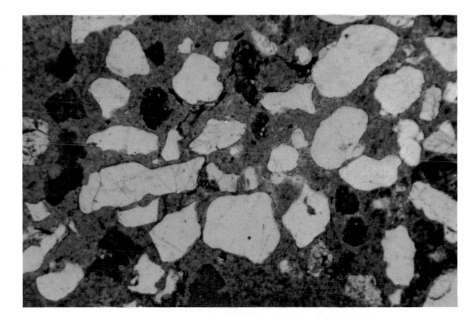

Lower Cretaceous Gault Fm.
England

A phosphate-cemented sandstone. Sand-sized grains of quartz and glauconite are encased in a pore-filling phosphatic cement which completely obliterated porosity. This cement has patchy, concretionary distribution concentrated within certain bedding-parallel intervals which appear to be horizons of especially slow sedimentation in a shallow marine environment.

0.30 mm

Lower Cretaceous Gault Fm.
England

Same as previous view but with crossed polarizers. Phosphate-cemented areas, which appeared as brownish material in the previous photo, appear isotropic under cross-polarized light. This apparent isotropism is a function of the very low birefringence of calcium fluorapatite coupled with very fine crystal size. Detrital quartz grains are abundant and numerous glauconite grains, characterized by a granular, greenish-brown, anomalous birefringence, can be seen.

XN   0.30 mm

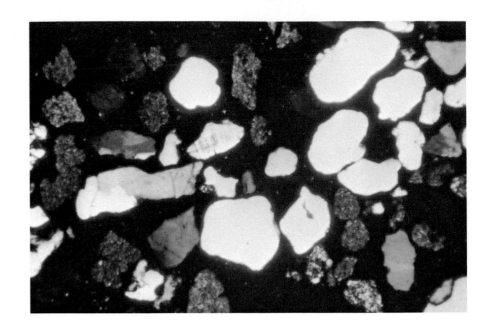

Pennsylvanian "Gray Ss." of Strawn Gp. ★
Texas ca. 1,370 m (4,500 ft)

A barite- and kaolinite-cemented sandstone. Kaolinite is the vermicular material with abundant intercrystalline porosity (pore space filled with blue plastic). Barite is the large, orthorhombic mineral with excellent cleavages and a refractive index higher than balsam. Barite commonly occurs as a cement in sandstones, as a replacement of limestones, or as hydrothermal, metalliferous veins. Photo by S. P. Dutton.

0.04 mm

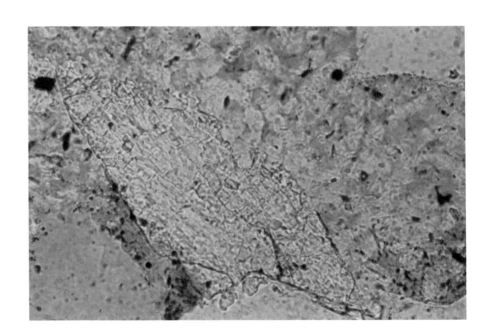

Mississippian limestone
Ireland

The entire photo shows barite crystals from a hydrothermal vein in limestone. The birefringence of barite is very near that of quartz but barite can be distinguished by its higher refractive index, biaxial figure, orthorhombic crystal form, and two perfect cleavages (with one additional poor cleavage). In hand sample, barite is also distinguishable by its high density.

XN   0.38 mm

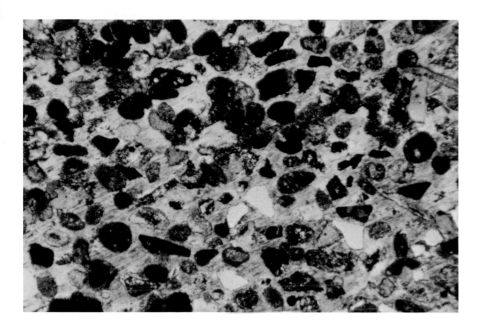

Holocene sabkha sands
Abu Dhabi

A gypsum-cemented sandstone. A single crystal of gypsum encompasses all the detrital carbonate and clastic grains in the field of view in a poikilotopic or "luster mottled" texture. This thin section is slightly thicker than the standard 30 μm so the gypsum shows unusually high birefringence colors (first-order gray is more typical). Although this cement obliterates porosity it has the potential to be leached in subsurface environments.

XN   0.30 mm

Pennsylvanian Tensleep Ss. *
Wyoming 5,253 m (17,236 ft)

Anhydrite-cemented sandstone. The anhydrite is characterized by third-order green birefringence colors, three distinct-to-perfect cleavages, and a slight etching of the surface (which causes the mottled appearance of the birefringence colors) produced during section grinding. As with gypsum and other evaporite cements, there exists significant potential for secondary porosity in anhydrite-cemented sands.

XN   0.10 mm

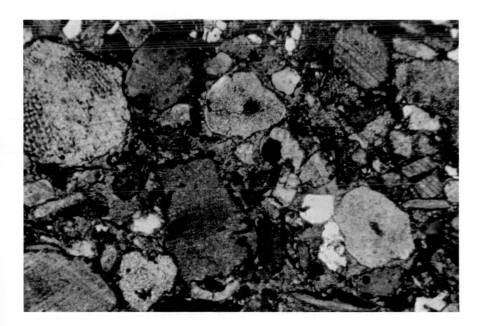

Upper Mississippian-Permian Nuka Fm.
Alaska

Glauconite-cemented calcareous sandstone. The grains with extreme birefringence are calcitic echinoderm fragments; quartz grains with low birefringence colors are also present. Greenish interstitial material is glauconite which is only rarely found as a cementing agent although it commonly fills small voids such as foraminiferal chambers. Glauconite cements are probably restricted to marine units.

XN   0.30 mm

Upper Cretaceous Kogosukruk Tongue of
  Prince Creek Fm.
Alaska

Clay minerals are a very important cementing agent in many sandstones. In this example, early stages of clay cementation are visible with thin, authigenic coatings of clays having formed around most of the detrital grains. Pore space is shown in bluish-green. Even such thin coatings may serve to isolate the grains from the pore fluids and thus inhibit alteration or cementation processes.

0.15 mm

Upper Cretaceous Kogosukruk Tongue of
  Prince Creek Fm.
Alaska

Close-up view of clay coatings from same sample as in previous illustration. Clays (probably mainly smectite-illite) form complete but thin coatings around detrital grains. Note brownish color and moderately high birefringence of clay films. Detrital grains in this example include chert and quartz.

XN   0.10 mm

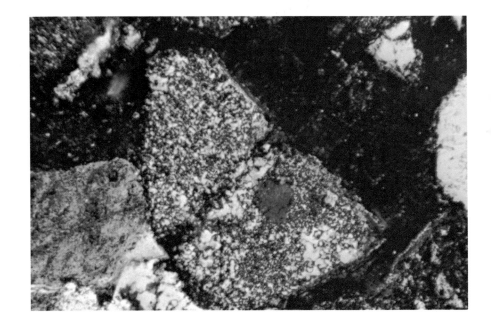

Upper Cretaceous Frontier Fm.
Wyoming ca. 610 m (2,000 ft)

SEM view of a smectite (montmorillonite) clay coating on a detrital quartz sand grain. In this case, the clays form a relatively dense and smooth coating (center of photo) on the detrital grain surface (lower left). Remnant porosity is visible in upper right. Photo by E. D. Pittman.

SEM   5 μm

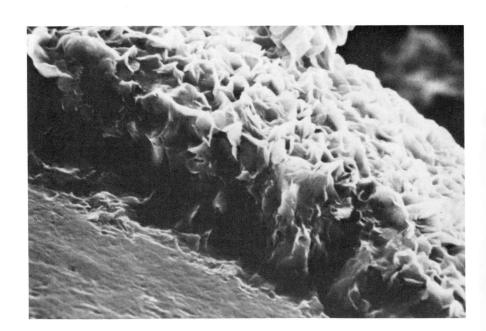

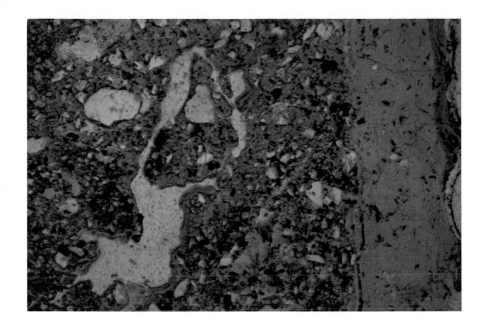

Quaternary soil
Australia

Clay minerals can, in some cases, form relatively early in the diagenetic history of sediments. Clays in this example include material which has infiltrated into the sediment as well as clays which have been neoformed. Podzolic soils such as this are subjected to intense eluviation which involves mechanical transport of clays as well as the neoformation of clay minerals from material carried in solution. Photo by E. F. McBride.

0.30 mm

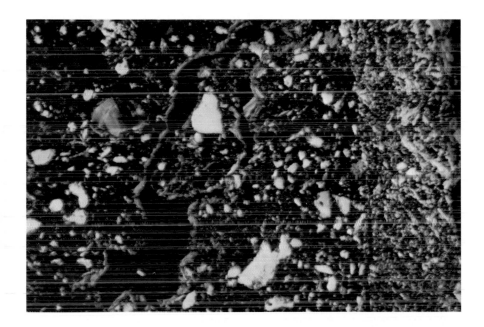

Quaternary soil
Australia

Same view as previous photo but with crossed polarizers. Neoformed clay minerals are visible, in association with iron oxides and hydroxides, filling small pore spaces throughout this soil, forming rims around detrital grains, and producing a thick crust at the right side of the photo. Photo by E. F. McBride.

XN   0.30 mm

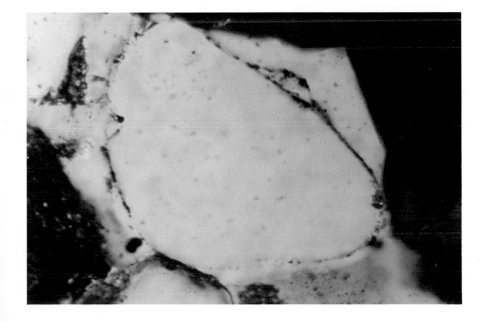

Lower Permian Brushy Canyon(?) Fm.
Texas

Relatively early diagenetic clay formation. Very thin clay rinds surround detrital quartz grains and are covered by subsequently formed quartz overgrowth cement and still later by diagenetic calcite. Clay coatings are presumably discontinuous because complete covering of the detrital quartz by clays would have prevented overgrowth formation. Although probably not detrital, these clay films may have formed during or right after deposition.

XN   0.025 mm

Upper Cretaceous Kogosukruk Tongue of ★
  Prince Creek Fm.
Alaska

Clay coatings of moderate thickness can
have a remarkably important effect in
reducing the permeability of sandstones.
Here, brownish crusts of authigenic clay
cements completely line pores and, in some
places, bridge pores. This may drastically
reduce pore throat sizes as well as effective
porosity and permeability.

0.025 mm

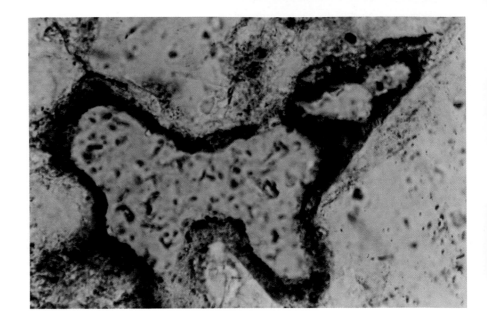

Lower Permian Rotliegendes Ss.
British North Sea 2,771 m (9,092 ft)

Authigenic clays (mainly illite) coating
detrital quartz grains. Wispy terminations
of clays extend into, and partially bridge,
pores. Although considerable porosity
remains, much of it is isolated by the
narrowness of the connecting pore throats.
Photo by E. D. Pittman.

SEM   22 μm

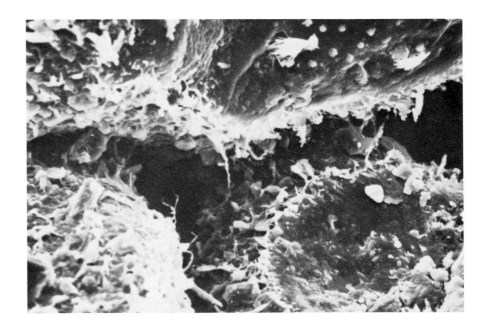

Lower Permian Rotliegendes Ss. ★
British North Sea ca. 1,980 m (6,500 ft)

Authigenic illite coatings on detrital sand
grains. In this example, the illite cement
forms a thick grain coating as well as
completely bridging pores in numerous
places. Fluid movements would be greatly
retarded by such cementation. Photo by
E. D. Pittman.

SEM   22 μm

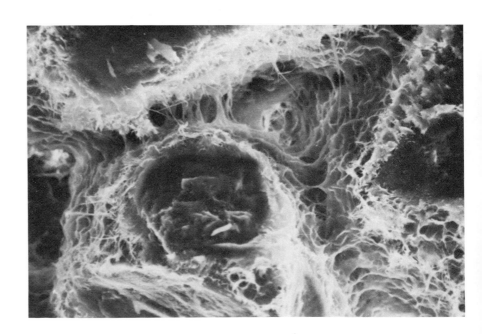

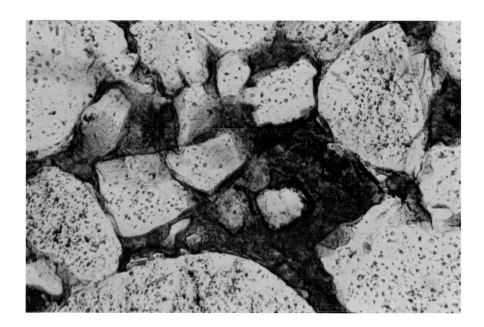

Upper Jurassic Salt Wash Mbr. of
  Morrison Fm.
Colorado

Clay cementation can proceed to the point of complete occlusion of porosity, as in this example. Diagenetic clay minerals can form from the alteration of precursor clays, from degradation of unstable detrital grains, or, possibly, they may be produced from component materials which have undergone long-distance transport in solution. In any case, an increase in overall grain volume may lead to significant overall porosity reduction.

0.06 mm

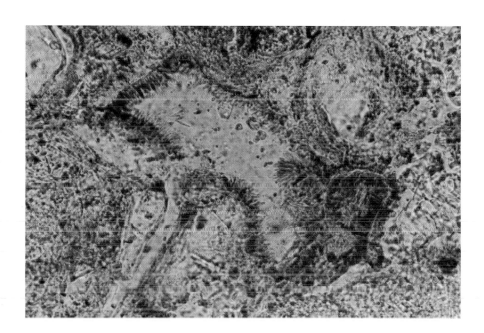

Miocene Cierbo and Neroloy Ss. ★
California

Radial, fibrous, clearly authigenic smectite (montmorillonite) clay films coating detrital sand grains. A pore-lining fabric of strongly oriented grains is an excellent indicator of an authigenic origin of clays. The presence of such clay films clearly has destroyed some porosity (shown in pale green) but it may also have prevented overgrowth cementation of quartz and other minerals.

0.04 mm

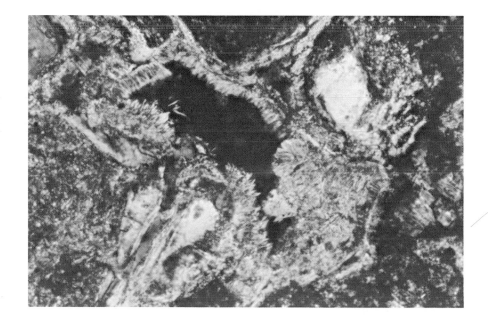

Miocene Cierbo and Neroloy Ss. ★
California

Same view as previous photo but with crossed polarizers. Note high birefringence of the authigenic, grain-coating clay films. X-ray diffraction analyses have shown that smectite is the major clay mineral in this rock; petrographic observations are necessary, however, to determine the authigenic nature of the clays and their distribution within the rock.

XN   0.04 mm

Upper Cretaceous Frontier Fm. ★
Wyoming ca. 610 m (2,000 ft)

Authigenic smectite (montmorillonite) as a cementing agent in a sandstone. The highly crenulate, honeycombed, interlocking crystals are typical of montmorillonite in SEM view. The smooth fusing of adjacent crystals serves to distinguish these crystals from chlorite. The high degree of crystallinity shown in these clays is a strong indicator of an authigenic origin; this criterion is valid for other clay minerals as well. Photo by E. D. Pittman.

SEM   3.3 μm

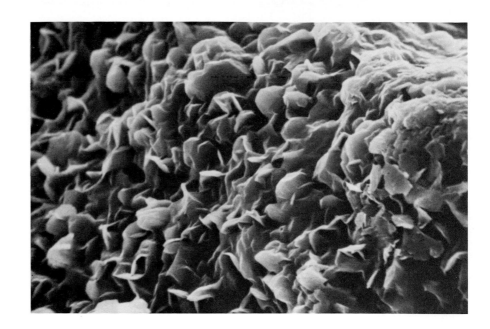

Lower Cretaceous Patula Arkose
Mexico

A chlorite cemented sandstone. Authigenic clay formation has extended beyond the "film" stage and has completely obliterated porosity. The light olive-green color is typical of chlorite. Under higher magnification the chlorite can be seen to occur as coarsely crystalline, fibrous crusts which line pores and are clearly authigenic.

0.10 mm

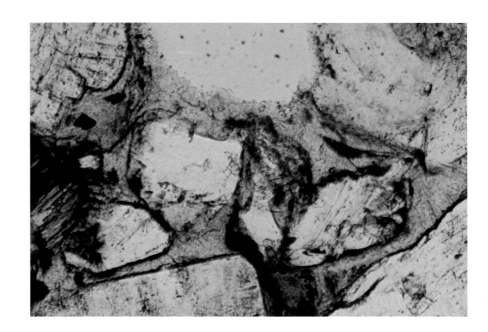

Lower Cretaceous Patula Arkose ★
Mexico

A different field of view from same sample as above. Using polarized light one can see the radial development of authigenic chlorite crusts. Note the "ultra-blue" anomalous birefringence colors characteristic of chlorite. Chlorite is an important cementing agent in many sandstones, especially in areas of deep burial.

XN   0.10 mm

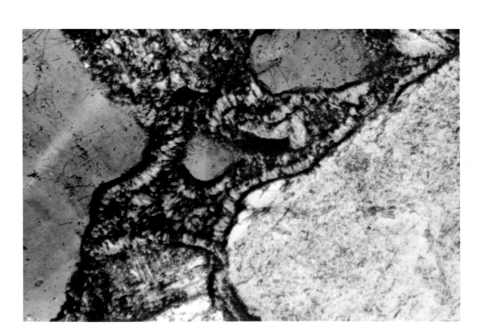

Upper Triassic New Haven Arkose
Connecticut

Authigenic vermicular chlorite. Chlorite (as
well as kaolinite and vermiculite) sometimes assumes this vermicular growth form
as "books" or "worms" of clay plates.
Note the "ultrablue" extinction colors
typical of chlorite. Some authigenic illite-
sericite is also visible in this sample.

XN   0.12 mm

Upper Cretaceous Tuscaloosa Fm. *
Louisiana 6,137 m (20, 136 ft)

Authigenic chlorite cement can have a wide
variety of crystal morphologies. In this
example, chlorite occurs as individual
idiomorphic crystals which are plate-like
and are attached to the detrital sand grains
along their thin edge. This is perhaps the
most common morphology for authigenic
chlorite. Photo by G. W. Smith.

SEM   3.5 μm

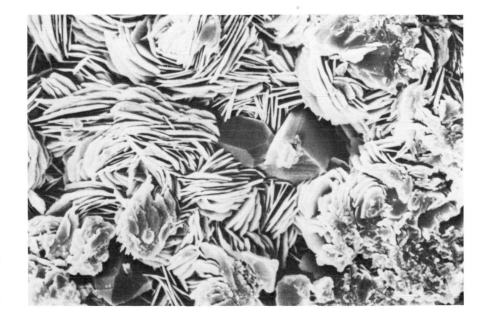

Oligocene Frio Fm.
Texas

Authigenic chlorite cement in the form of
rosette-shaped clusters associated with
larger authigenic quartz crystal. This common growth morphology can form either
as a pore lining or pore filling texture. The
individual crystals are characterized by
having smooth but lobate edges which
helps to differentiate them from super-
ficially similar kaolinite or cristobalite
spherules. Photo by E. D. Pittman.

SEM   4 μm

Upper Triassic New Haven Arkose
Connecticut

Λ dominantly illite-sericite cemented sandstone. The clays in this example consist of
a complex mixture of some kaolinite and
chlorite plus very abundant illite-sericite.
Some of the clay is detrital matrix, some is
produced by *in situ* alteration of unstable
minerals, and some is clearly authigenic
cement. Defining exact amounts of each of
these categories is most difficult. Illite-
sericite is recognized here by is relatively
high birefringence and coarse texture.

XN   0.15 mm

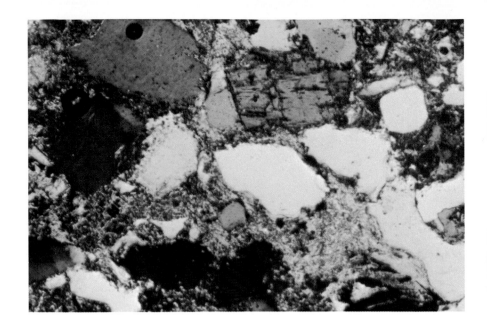

Lower Permian Rotliegendes Ss. ★
British North Sea ca. 1,980 m (6,500 ft)

SEM view of authigenic illite cement in a
sandstone. This delicate growth form,
consisting of sheets with wispy to latch-
like terminations, is the most common
habit for illitic cements in sandstones. Pure
illites can be confused with mixed-layer
montmorillonite-illite.   Identifications
should, therefore, be checked with X-ray
diffraction. Photo by E. D. Pittman.

SEM   7 µm

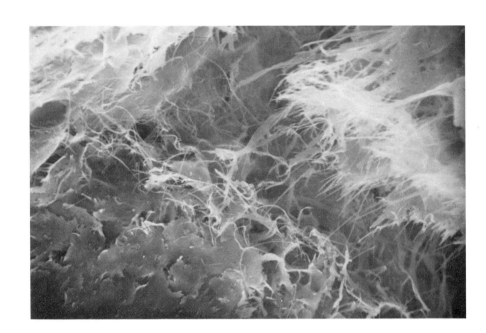

Lower Cretaceous Muddy Ss.
Wyoming

Kaolinite cemented sandstone. Kaolinite
(with small amounts of illite) formed at
least partly authigenically, completely
obliterating porosity in this sample. Brownish color is produced by optical effects
related to the overlap of numerous small
clay flakes and by minute inclusions within
the clays.

0.04 mm

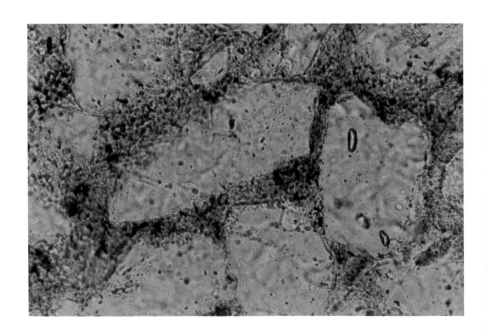

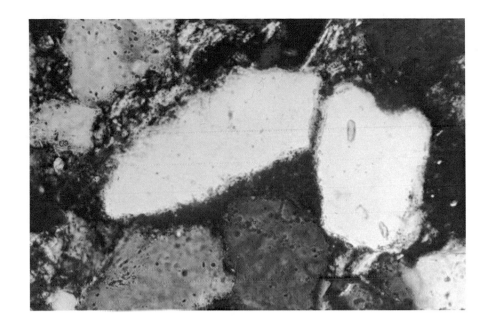

Lower Cretaceous Muddy Ss.
Wyoming

Same view as previous photo but with crossed polarizers. Kaolinite cement shows characteristic low birefringence. The coarser, higher birefringent clay minerals in this example are mainly illite. Kaolinite is a very common cementing agent in sandstones, although it commonly only partially fills pore spaces.

XN   0.04 mm

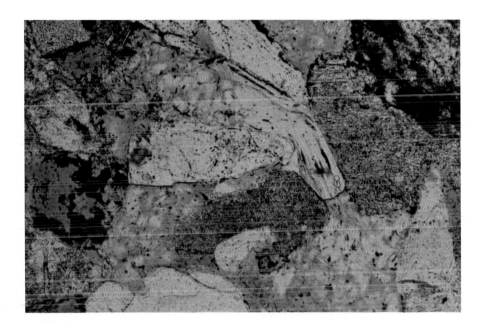

Triassic Dockum Gp. ★
Texas

Vermicular kaolinite-dickite filling pore space (marked by blue-stained plastic) in a sandstone. Individual vermicular stacks commonly completely bridge pores and sharply reduce overall permeability. Porosity, however, is rarely completely obliterated as is evident from this example showing extensive remnant intercrystalline porosity. Note deformed muscovite flake (top center) and leached grain (left edge).

0.05 mm

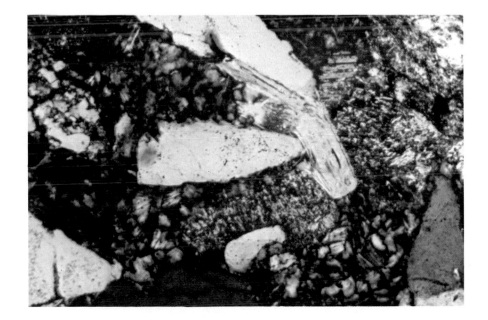

Triassic Dockum Gp. ★
Texas

Same view as previous photo but with crossed polarizers. Kaolinite is readily identified by its vermicular texture in this example, coupled with its low birefringence. Kaolinite is very difficult to distinguish from dickite using either light microscopy or SEM; the minerals can be readily distinguished using X-ray diffraction, however.

XN   0.05 mm

138

Eocene Huerfano Fm.
Colorado

Vermicular kaolinite encased in a calcite cement crystal. In this example, vermicular stacks of kaolinite partially filled pore space in a sandstone. Subsequent formation of coarsely crystalline calcite cement led to the incorporation of the highly porous kaolinite within the calcite.

XN   0.05 mm

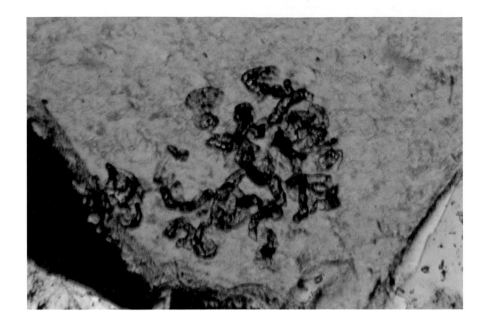

Upper Cretaceous Tuscaloosa Fm. *
Louisiana 4,923 m (16,150 ft)

Kaolinite and dickite are characterized in SEM by vermicular stacks of pseudohexagonal plates and thus are, perhaps, the most readily recognizable of all the clays. The vermicular kaolinite stacks seen here are generally curved and intertwining, with considerable intercrystalline porosity (this is presumably associated with very low intercrystalline permeability, however, because of the small pore diameters). Photo by G. W. Smith.

SEM   5 μm

Tertiary 'Vieja Group' *
Texas

A zeolite cemented sandstone. Radiating clusters of needle-like crystals of mordenite(?) have cemented and partially replaced this volcaniclastic sediment. Zeolites have a wide range of compositions, textures, and optical properties. Because of their generally fine grain size, and low birefringence they are commonly overlooked or misidentified. Nevertheless, zeolites can be very important cementing agents, especially in volcaniclastic sediments.

XN   0.38 mm

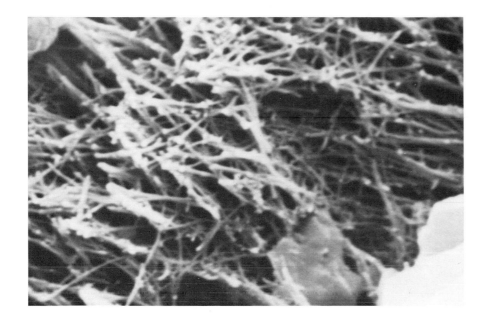

Tertiary sediment
Oregon

Mordenite in SEM view. The individual, interlocking, oriented fibrous crystals are clearly visible. Most zeolites, including mordenite, are commonly associated with volcanic material (especially vitric rhyolite ash). They also occur frequently in ash-rich, restricted, alkaline lake beds and as burial-diagenetic or low-grade metamorphic minerals. Photo by A. J. Gude, III.

SEM   1.3 μm

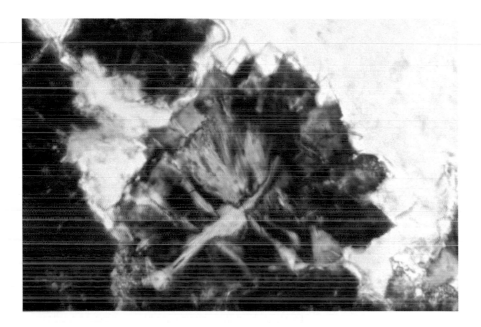

Holocene and Pleistocene Honolulu
  Volcanics
Oahu, Hawaii

Phillipsite spherule surrounded by calcite in a palagonite tuff. Phillipsite, a monoclinic zeolite, has optical properties similar to most other common sedimentary zeolites—very low birefringence, low index of refraction, and very fine crystal size. Phillipsite has been recognized in lacustrine sediments as well as in deep marine sections of slowly deposited, volcanically derived sediments.

XN   0.025 mm

Pliocene Big Sandy Fm. *
Arizona

Phillipsite spherule in SEM. This example is from tuffaceous sediments in an arid non-marine environment (fluviatile and lacustrine). Although phillipsite is commonly spherulitic, it may also be found as a pseudomorph after shards, as prismatic crystals, or as fibers. Photo by A. J. Gude, III.

SEM   5 μm

140

Pleistocene sediment, Lake Tecopa
California

Erionite crystals in an altered rhyolitic vitric tuff. Erionite (a zeolite) typically has a prismatic or acicular habit but here is seen as oriented bundles of radiating crystals. Erionite has a low refractive index and a relatively higher birefringence for a zeolite (although its birefringence is still considerably lower than that of quartz).

XN   0.06 mm

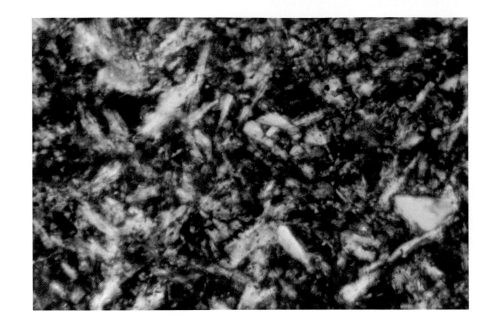

Pleistocene sediment, Lake Tecopa
California

Erionite crystals in SEM. Orthorhombic, striated, partly intergrown, needle-like crystals are typical of erionite morphology. This sample is from altered tuffaceous sediments deposited in an alkaline lake. Photo by A. J. Gude, III.

SEM   5 μm

Tertiary ash flow tuff
New Mexico

Relatively large prismatic to platy crystals of clinoptilolite line molds of former glass shards preserving the relict vitroclastic texture. Clinoptilolite, a zeolite of the heulandite structural group, is one of the most common zeolites found in sedimentary rocks. It has been described from fluvial, lacustrine, and marine settings, and in rocks as old as Jurassic. Low birefringence and low refractive index are characteristic.

XN   0.06 mm

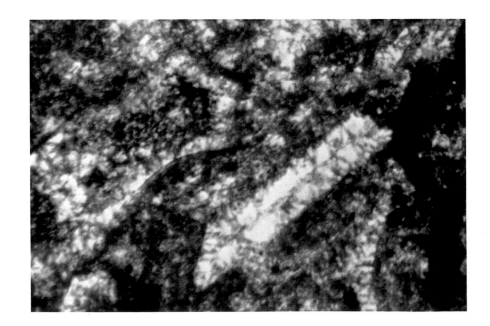

Miocene Barstow Fm. ★
California

Clinoptilolite in SEM view. Although clinoptilolite can be identified using a combination of standard petrography and SEM examination, X-ray analysis is the most certain method of identification. In SEM, clinoptilolite commonly appears as jumbled stacks of platy crystals, as in this example from altered tuffaceous sediments. Photo by A. J. Gude, III.

SEM   5 μm

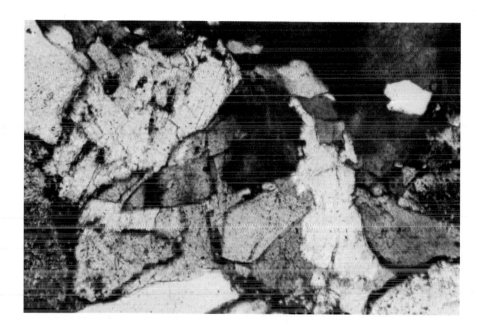

Paleocene Poison Canyon Fm.
Colorado

A laumontite cemented arkosic sandstone. Laumontite is one of the few zeolites which is not restricted to tuffaceous sedimentary rocks. In this example, laumontite occurs both as a coarse, pore-filling cement and as a replacement of alkali feldspars. It is recognized by its yellowish color with crossed polarizers (birefringence slightly higher than quartz) and by the common presence of two cleavages. It is an important burial-diagenetic cementing agent in many areas.

XN   0.10 mm

Pleistocene sediment, Lake Tecopa
California

Searlesite filling large voids and pseudomorphic glass shards in a tuffaceous sediment. Searlesite, a monoclinic zeolite, has low birefringence (similar to that of quartz), commonly has prismatic crystals which show good 100 cleavage, and normally forms relatively late in the paragenetic sequence of zeolites in tuffs.

XN   0.10 mm

Pliocene Big Sandy Fm.
Arizona

Chabazite, a well crystalized zeolite which is moderately common in upper Tertiary lacustrine tuffaceous units. A hexagonal mineral, chabazite commonly has euhedral rhombohedral crystals which make it quite distinctive from most other zeolites in SEM. Photo by A. J. Gude, III.

SEM   3 μm

Pliocene Big Sandy Fm.
Arizona

Analcime (analcite) is one of the most abundant zeolites in sedimentary rocks, especially in older sedimentary units. It is one of the few zeolites which occurs in rocks lacking vitric material as well as in tuffaceous rocks. Analcime commonly has euhedral to subhedral crystals of moderate size. Its birefringence is extremely low to isotropic. It is most easily identified using X-ray or SEM techniques.

SEM   12 μm

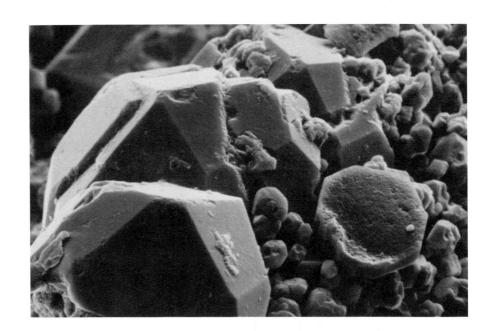

Miocene and Pliocene Monterey Fm.
California

Hydrocarbons such as oil and aphaltic residues may also effectively fill pore space in sandstones. In this example of a tar sand, amorphous, brownish-to-black, semi-solid petroleum hydrocarbons line and extensively fill pore space. Distinction of hydrocarbons from clays and other fine-grained cements is sometimes facilitated by fluorescence with ultraviolet light (for "live" oil) or by extraction with solvents.

0.22 mm

Pennsylvanian Strawn Gp. ★
Texas

Another example of pore-filling hydrocarbons, in this case, largely "dead" oil. Note bridging of pore space in some areas by thick films of oil. Distinction of hydrocarbons from clays can be quite difficult where "dead" oil is involved. Hydrocarbon cementation can be very effective in reducing permeability, especially where bridging of pores is prevalent.

0.27 mm

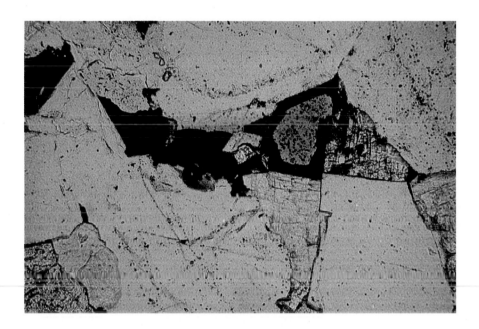

Cambrian Old Fort Island Ss.
Canada (N.W.T.)

Complex cementation of sandstone. Quartz overgrowth cementation was followed by complete porosity occlusion by dolomite formation. Subsequent partial dissolution of dolomite yielded secondary porosity which was partly filled with bitumen. Thus, in this example, petrographic analysis establishes the time of oil migration as being subsequent to quartz-dolomite cementation and secondary porosity formation. Photo by D. A. McDonald.

0. 15 mm

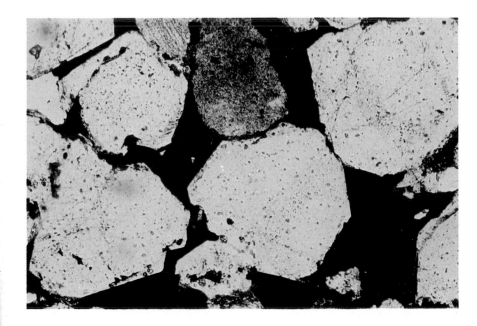

Pennsylvanian Ricker Ss. Mbr. of
 Mineral Wells Fm.
Texas

The establishment of paragenetic sequences of cementation, deformation, dissolution, and other diagenetic factors is one of the most important products of petrographic analysis. Here, quartz overgrowths were the earliest cementation event and were followed by complete porosity destruction through hematite formation. Quartz overgrowths are recognizable primarily because of their euhedral crystal terminations.

0.10 mm

Lower Permian Brushy Canyon Fm.
Texas

Complex cementation of sandstone. Quartz overgrowths formed as the first generation of cement and were followed by calcite which both filled pore space and marginally replaced the quartz overgrowths. Such textural relationships can be determined with relatively little expense in time and effort using petrography.

XN 0.06 mm

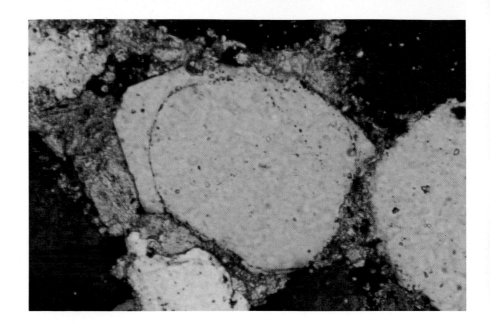

Upper Cretaceous Frontier Fm.
Wyoming ca. 610 m (2,000 ft)

Multiple stages of cementation and their relative timing can also be determined using SEM techniques. In this example, montmorillonite coats detrital grains and is followed by a later generation of kaolinite cement (upper part of photo). Photo by E. D. Pittman.

SEM 7 μm

Upper Cretaceous Tuscaloosa Fm. ★
Louisiana 5,073 m (16,645 ft)

An example of multiple generations of cementation visible using SEM. Detrital grains are completely coated with a rind of radially oriented, platy chlorite crystals. These are succeeded by a second generation of kaolinite (or dickite) cement in the form of short, vermicular stacks of pseudohexagonal crystals. Considerable remnant, intercrystalline porosity can still be seen. Photo by G. W. Smith.

SEM 12 μm

Selected Diagenesis-Porosity Bibliography

General

Adams, W. L., 1964, Diagenetic aspects of lower Morrowan, Pennsylvanian, sandstones, northwestern Oklahoma: AAPG Bull., v. 48, p. 1568-1580.

Ballance, P. F., and C. S. Nelson, 1969, Differential cementation in the Waikawau limestone (Waitemata Group). West Auckland, New Zealand: New Zeland Jour. Geol. Geophys., v. 12, p. 67-86.

Beard, D. C., and P. K. Weyl, 1973, Influence of texture on porosity and permeability of unconsolidated sand: AAPG Bull., v. 57, no. 3, p. 349-369.

Berg, R. R., 1975, Capillary pressures in stratigraphic traps: AAPG Bull., v. 59, p. 939-956.

Blatt, Harvey, 1966, Diagenesis of sandstones: Processes and problems, in J. R. Smith, ed., Symposium on recently developed geologic principles and sedimentation of the Permo-Pennsylvanian of the Rocky Mountains: Wyoming Geol. Assoc. 20th Field Conf. Guidebook, p. 65a-65o.

Choquette, P. W., and L. C. Pray, 1970, Geologic nomenclature and classification of porosity in sedimentary carbonates: AAPG Bull., v. 54, p. 207-250.

Dapples, E. C., 1962, Stages of diagenesis in the development of sandstones: Geol. Soc. America Bull., v. 73, p. 913-934.

———— 1967, Diagenesis of sandstones, in G. Larsen, and G. V. Chilingar, eds., Diagenesis in sediments, Devel. in sedimentology, Volume 8: New York, Elsevier, p. 91-125.

———— 1972, Some concepts of cementation and lithification of sandstones: AAPG Bull., v. 56, p. 3-25.

Dutton, S. P., 1977, Diagenesis and porosity distribution in deltaic sandstone, Strawn Series (Pennsylvanian), north-central Texas: Gulf Coast Assoc. Geol. Socs., Trans., v. 27, p. 272-277.

Emery, K. O., and S. C. Rittenberg, 1952, Early diagenesis of California basin sediments in relation to origin of oil: AAPG Bull., v. 36, p. 735-806.

Esch, H., 1966, Vergleichende diagenese-studien an Sandsteinen und Schiefertonen des Oberkarbon in Nordwestdeutschland und den East Midland in England: Fortschr. Geol. Rheinl. Westf., v. 10, p. 1013-1084.

Fairbridge, R. W., 1967, Phases of diagenesis and authigenesis, in G. Larsen and G. V. Chilingar, eds., Developments in Sedimentology, Volume 8: Diagenesis in Sediments: New York, Elsevier, p. 19-90.

Fothergill, C. A., 1955, The cementation of oil reservoir sands and its origin: 4th World Petrol. Cong. Proc., Sec. 1B, Paper 1, p. 301-314.

Füchtbauer, Hans, 1967, Influence of different types of diagenesis on sandstone porosity: 7th World Petrol. Cong., Proc., v. 2, p. 353-369.

———— 1974, Some problems of diagenesis in sandstones, in Sedimentation argilo-sableuse et diagenese: Centre Rech. Pau Bull., v. 8, p. 391-403.

Gaither, A., 1953, A study of porosity and grain relationships in experimental sands: Jour. Sed. Petrology, v. 23, p. 180-195.

Gilbert, C. M., 1949, Cementation of some California Tertiary reservoir sands: Jour. Geology, v. 57, p. 1-17.

Glover, J. E., 1963, Studies in the diagenesis of some western Australian sedimentary rocks: Jour. Royal Soc. Western Australia, v. 46, p. 33-56.

Heald, M. T., 1956a, Cementation of Simpson and St. Peter Sandstones in parts of Oklahoma, Arkansas, and Missouri: Jour. Geology, v. 64, p. 16-30.

———— 1956b, Cementation of Triassic arkoses in Connecticut and Massachusetts: Geol. Soc. America Bull., v. 67, p. 1133-1154.

———— 1965, Lithification of sandstones in West Virginia: West Virginia Geol. Survey Bull., v. 30, p. 1-28.

———— and J. J. Renton, 1966, Experimental study of sandstone cementation: Jour. Sed. Petrology, v. 36, p. 977-991.

Horn, D., 1965, Diagenese und porosität des Dogger-beta-Hauptsandsteines in den Öfeldern Plon mit Preets: Erdöl Kohle, v. 18, p. 249-255.

Hsü, K. J., 1977, Studies of Ventura field, California, I: Lithology, compaction, and permeability of sands: AAPG Bull., v. 61, p. 137-168.

Jacka, A. D., 1970, Principles of cementation and porosity-occlusion in Upper Cretaceous sandstones, Rocky Mountain region: 22nd Ann. Field Conf., Guidebook, Wyoming Geol. Assoc., p. 265-285.

Jonas, E., and E. F. McBride, 1977, Diagenesis of sandstone and shale: application to exploration for hydrocarbons: Austin, Texas, Dept. of Geol. Sciences, Univ. of Texas at Austin, Continuing Education Program Pub. No. 1, 167 p.

Kieke, E. M., and D. J. Hartmann, 1973, Scanning electron microscope application to formation evaluation: Gulf Coast Assoc. Geol. Socs., Trans., v. 23, p. 60-67.

———— 1974, Detecting microporosity to improve formation evaluation: Jour. Petroleum Technology, v. 26, p. 1080-1086.

Kossovskaya, A. G., ed., 1970, Lithification of clastic sediments: Sedimentology, v. 15, p. 7-204.

Krynine, P. D., 1941, Petrographic studies of variations in cementing material in the Oriskany Sand: Proc. 10th Penna. Min. Industries Conf. Penna. State College Bull., v. 33, p. 108-116.

Levandowski, D. W., M. E. Kaley, S. R. Silverman, and R. G. Smalley, 1973, Cementation in Lyons Sandstone and its role in oil accumulation, Denver Basin, Colorado: AAPG Bull., v. 57, p. 2217-2244.

Lindquist, S. J., 1977, Secondary porosity development and subsequent reduction, overpressured Frio Formation sandstone (Oligocene), south Texas: Gulf Coast Assoc. Geol. Socs., Trans., v. 27, p. 99-107.

Loucks, R. G., D. G. Bebout, and W. E. Galloway, 1977, Relationship of porosity formation and preservation to sandstone consolidation history—Gulf Coast Lower Tertiary Frio Formation: Gulf Coast Assoc. Geol. Socs., Trans., v. 27, p. 109-120.

Maxwell, J. C., 1960, Experiments on compaction and cementation of sand, in D. Griggs and J. Handin, eds., Rock deformation: Geol. Soc. America Mem. 79, p. 105-132.

———— 1964, Influence of depth, temperature, and geologic age on porosity of quartzose sandstones: AAPG Bull., v. 48, p. 697-709.

McBride, E. F., 1977, Secondary porosity—importance in sandstone reservoirs in Texas: Gulf Coast Assoc. Geol. Socs., Trans., v. 27, p. 121-122.

Morgan, J. T., and D. T. Gordon, 1970, Influence of pore geometry on water-oil relative permeability: Jour. Petrol. Technol., v. 22, p. 1199-1208.

Morrow, N. R., J. D. Huppler, and A. B. Simmons, III, 1969, Porosity and permeability of unconsolidated, Upper Miocene sands from grain-size analysis: Jour. Sed. Petrology, v. 39, p. 312-321.

Packham, G. H., and K. A. W. Crook, 1960, The principle of diagenetic facies and some of its implications: Jour. Geology, v. 68, p. 392-407.

Parker, C. A., 1974, Geopressure and secondary porosity in the deep Jurassic of Mississippi: Gulf Coast Assoc. Geol. Socs., Trans., v. 24, p. 69-80.

Pittman, E. D., and R. W. Duschatko, 1970, Use of pore casts and scanning electron microscope to study pore geometry: Jour. Sed. Petrology, v. 40, p. 1153-1157.

Prozorovich, G. E., 1970, Determination of the time of oil and gas accumulation by epigenesis studies: Sedimentology, v. 15, p. 41-52.

Riezbos, P. A., 1974, Scanning electron microscopical observations on weakly cemented Miocene sands: Geologie en Mijnbouw, v. 53, p. 109-122.

Rittenhouse, Gordon, 1971, Pore-space reduction by solution and cementation: AAPG Bull., v. 55, p. 80-91.

——— 1973, Pore-space reduction in sandstones—controlling factors and some engineering implications: Offshore Tech. Conf. Prepr. No. 5, v. 1, p. 683-688.

Sarkisyan, S. G., 1971, Application of the scanning electron microscope in the investigation of oil and gas reservoir rocks: Jour. Sed. Petrology, v. 4, p. 289-292.

Schmidt, V., D. A. McDonald, and R. L. Platt, 1977, Pore geometry and reservoir aspects of secondary porosity in sandstones: Bull. Canadian Petrol. Geol., v. 25, p. 271-290.

Sippel, R. F., 1968, Sandstone petrology, evidence from luminescence petrography: Jour. Sed. Petrology, v. 38, p. 530-554.

Stanton, G. D., 1977, Secondary porosity in sandstones of the Lower Wilcox (Eocene), Karnes County, Texas: Gulf Coast Assoc. Geol. Socs., Trans., v. 27, p. 197-207.

Sujkowski, Z. L., 1958, Diagenesis: AAPG Bull., v. 42, p. 2692-2727.

Tallman, S. L., 1949, Sandstone types: Their abundance and cementing agents: Jour. Geology, v. 57, p. 582-591.

Taylor, J. M., 1950, Pore-space reduction in sandstones: AAPG Bull., v. 34, p. 701-716.

Von Engelhardt, W., 1967, Interstitial solution and diagenesis in sediments, in G. Larsen and G. V. Chilinger, eds., Developments in Sedimentology Volume 8: Diagenesis in Sediments: New York, Elsevier, p. 503-522.

Waldschmidt, W. A., 1941, Cementing materials in sandstones and their probable influence on the migration and accumulation of oil and gas: AAPG Bull., v. 25, p. 1839-1879.

Warner, M. M., 1965, Cementation as a clue to structure drainage patterns, permeability, and other factors: Jour. Sed. Petrology, v. 35, no. 4, p. 797-804.

Weinbrandt, R. M., and Irving Fatt, 1969, A scanning electron microscope study of the pore structure of sandstone: Jour. Petroleum Tech., v. 21, p. 543-548.

Yurkova, R. M., 1970, Comparison of post-sedimentary alteration of oil-, gas-, and water-bearing rocks: Sedimentology, v. 15, p. 53-68.

Silica and Quartz

Cecil, C. B., and M. T. Heald, 1971, Experimental investigation of the effects of grain coatings on quartz growth: Jour. Sed. Petrology, v. 41, p. 582-584.

Cook, P. J., 1970, Repeated diagenetic calcitization, phosphatization, and silicification in the Phosphoria Formation: Geol. Soc. America Bull., v. 81, p. 2107-2116.

Dapples, E. C., 1959, The behavior of silica in diagenesis, in H. A. Ireland, ed., Silica in sediments: Soc. Econ. Paleontologists and Mineralogists Spec. Pub. No. 7, p. 36-51.

——— 1967, Silica as an agent in diagenesis, in G. Larsen and G. V. Chilingar, eds., Developments in sedimentology, Volume 8: Diagenesis in sediments: New York, Elsevier, p. 323-342.

Deelman, J. C., 1976, "Dust rings" or vacancy clouds in quartz grains: Naturwissenschaften, v. 63, p. 37.

Dietrich, R. V., C. R. R. Hobbs, Jr., and W. D. Lowry, 1963, Dolomitization interrupted by silicification: Jour. Sed. Petrology, v. 33, p. 646-663.

Ernst, W. G., and H. Blatt, 1964, Experimental study of quartz overgrowths and synthetic quartzites: Jour. Geology, v. 72, p. 461-470.

Fairbridge, R. W., 1975, Epidiagenetic silicification: 9th Cong. Internat. Sedimentologie, Nice, theme 7, p. 49-54.

Folk, R. L., and J. S. Pittman, 1971, Length-slow chalcedony: A new testament for vanished evaporites: Jour. Sed. Petrology, v. 41, p. 1045-1058.

Frye, J. C., and A. Swinford, 1946, Silicified rock in the Ogallala Formation: Kansas Geol. Survey Bull., no. 64, pt. 2, p. 37-76.

Greenwood, Robert, 1973, Cristobalite: Its relationship to chert formation in selected samples from the Deep Sea Drilling Project: Jour. Sed. Petrology, v. 43, p. 700-708.

Heald, M. T., and R. C. Anderegg, 1960, Differential cementation in the Tuscarora Sandstone: Jour. Sed. Petrology, v. 30, p. 568-577.

——— and R. E. Larese, 1974, Influence of coatings on quartz cementation: Jour. Sed. Petrology, v. 44, p. 1269-1274.

Humphries, D. W., 1956, Chert: Its age and origin in the Hythe Beds of the western Wealden: Geologists Assoc. London, Proc., v. 67, p. 296-313.

Krauskopf, K. B., 1956, Dissolution and precipitation of silica at low temperatures: Geochim. et Cosmochim. Acta, v. 10, p. 1-26.

Lancelot, Y., 1973, Chert and silica diagenesis in sediments from the central Pacific, in Initial Reports of the Deep Sea Drilling Project: Washington, D.C., U.S. Govt. Printing Office, v. 17, p. 377-405.

Lerbekmo, J. F., and R. L. Platt, 1962, Promotion of pressure solution of silica in sandstones: Jour. Sed. Petrology, v. 32, p. 514-519.

Mackenzie, F. T., and R. Gees, 1971, Quartz: Synthesis at earthsurface conditions: Science, v. 173, p. 533-535.

McBride, E. F., W. L. Lindemann, and P. S. Freeman, 1968, Lithology and petrology of the Gueydan (Catahoula) Formation in south Texas: Univ. Texas Bur. Econ. Geol. Rept. Inv. 63, 122 p.

Mitsui, Kiyohiro, and K. Taguchi, 1977, Silica mineral diagenesis in Neogene Tertiary shales in the Tempoku District, Hokkaido, Japan: Jour. Sed. Petrology, v. 47, p. 158-167.

Mitzutani, S., 1970, Silica minerals in the early stage of diagenesis: Sedimentology, v. 15, p. 419-436.

Oehler, J. H., 1975, Origin and distribution of silica lepispheres in porcelanite from the Monterey Formation of California: Jour. Sed. Petrology, v. 45, p. 252-257.

Paraguassu, A. B., 1972, Experimental silicification of sandstone: Geol. Soc. America Bull., v. 83, p. 2853-2858.

Pittman, E. D., 1972, Diagenesis of quartz in sandstones as revealed by scanning electron microscopy: Jour. Sed. Petrology, v. 42, p. 507-519.

Pittman, J. S., Jr., 1959, Silica in Edwards Limestone, Travis County, Texas, in H. A. Ireland, ed., Silica in sediments: Soc. Econ. Paleontologists and Mineralogists Spec. Pub. No. 7, p. 121-134.

Senior, B. R., and D. A. Senior, 1972, Silcrete in southwest Queensland: Australia Bur. Mineral Resources Geology and Geophysics Bull. No. 125, p. 23-28.

Siever, R., 1959, Petrology and geochemistry of silica cementation in some Pennsylvanian sandstones, in H. A. Ireland, ed., Silica in sediments: Soc. Econ. Paleontologists and Mineralogists Spec. Pub. No. 7, p. 55-79.

——— 1962, Silica solubility, 0°-200° C, and the diagenesis of siliceous sediments: Jour. Geology, v. 70, p. 127-150.

Sippel, R. F., 1968, Sandstone petrology, evidence from luminescence petrography: Jour. Sed. Petrology, v. 38, p. 530-554.

Swineford, A., and P. C. Franks, 1959, Opal in the Ogallala Formation in Kansas, in H. A. Ireland, ed., Silica in sediments: Soc. Econ. Paleontologists and Mineralogists Spec. Pub. No. 7, p. 111-120.

Towe, K. M., 1962, Clay mineral diagenesis as a possible source of silica cement in sedimentary rocks: Jour. Sed. Petrology, v. 32, p. 26-28.

Walker, T. R., 1960, Carbonate replacement of detrital crystalline silicate minerals as a source of authigenic silica in sedimentary rocks: Geol. Soc. America Bull., v. 91, p. 145-152.

—— 1962, Reversible nature of chert-carbonate replacement in sedimentary rocks: Geol. Soc. America Bull., v. 73, p. 237-242.

Wallace, C. A., 1976, Diagenetic replacement of feldspar by quartz in the Uinta Mountain Group, Utah and its geochemical implications: Jour. Sed. Petrology, v. 46, p. 847-861.

Waugh, B., 1970a, Formation of quartz overgrowths in the Penrith Sandstone (Lower Permian) of northwest England as revealed by scanning electron microscopy: Sedimentology, v. 14, p. 309-320.

—— 1970b, Petrology, provenance and silica diagenesis of the Penrith Sandstone (Lower Permian) of northwest England: Jour. Sed. Petrology, v. 40, p. 1226-1240.

Williamson, W. O., 1957, Silicified sedimentary rocks in Australia: Am. Jour. Sci., v. 255, p. 23-42.

Wilson, R. C. L., 1966, Silica diagenesis in Upper Jurassic Limestones of southern England: Jour. Sed. Petrology, v. 36, p. 1036-1049.

Wise, S. W., Jr., R. F. Buie, and F. M. Weaver, 1972, Chemically precipitated sedimentary cristobalite and the origin of chert: Eclogae geol. Helvet., v. 65, p. 157-163.

Carbonates

Allen, R. C., G. Eliezer, G. M. Friedman, and J. E. Sanders, 1969, Aragonite-cemented sandstone from outer continental shelf off Delaware Bay: Submarine lithification mechanism yields product resembling beachrock: Jour. Sed. Petrology, v. 39, p. 136-150.

Bogoch, R., and P. Cook, 1974, Calcite cementation of a Quaternary conglomerate in southern Sinai: Jour. Sed. Petrology, v. 44, p. 917-920.

Bricker, O. P., ed., 1971, Carbonate cements: Baltimore, Johns Hopkins Press, 376 p.

Dapples, E. C., 1971, Physical classification of carbonate cement in quartzose sandstones: Jour. Sed. Petrology, v. 41, p. 196-204.

Davies, D. K., 1967, Origin of friable sandstone; Calcareous sandstone rhythms in the upper Lias of England: Jour. Sed. Petrology, v. 37, p. 1179-1188.

Deffeyes, K. S., F. J. Lucia, and P. K. Weyl, 1965, Dolomitization of Recent and Plio-Pleistocene sediments by marine evaporite waters on Bonaire, Netherlands Antilles, in L. C. Pray, and R. C. Murray, eds., Dolomitization and limestone diagenesis, a symposium: Soc. Econ. Paleontologists and Mineralogists Spec. Pub. 13, p. 71-88.

Donovan, T. J., I. Friedman, and J. D. Gleason, 1974, Recognition of petroleum-bearing traps by unusual isotopic compositions of carbonate-cemented surface rocks: Geology, v. 2, no. 7, p. 351-354.

Evamy, B. D., and D. J. Shearman, 1965, The development of overgrowths from echinoderm fragments: Sedimentology, v. 5, p. 211-233.

Folk, R. L., 1974, The natural history of crystalline calcium carbonate: Effects of magnesium content and salinity: Jour. Sed. Petrology, v. 40, p. 40-53.

—— and L. S. Land, 1975, Mg/Ca ratio and salinity: Two controls over crystallization of dolomite: AAPG Bull., v. 59, p. 60-68.

Friedman, G. M., 1968, The fabric of carbonate cement and matrix and its dependence on the salinity of water, in G. Müller and G. M. Friedman, eds., Recent developments in carbonate sedimentology in central Europe: New York, Springer Verlag, p. 11-20.

Fuhrmann, W., 1968, "Sandkristalle" und Kugelsandsteine. Ihre Rolle bei der Diagenese von Sanden: Der Aufschluss, v. 5, p. 105-111.

Garrison, R. E., J. L. Luternauer, E. V. Grill, R. D. Macdonald, and J. W. Murray, 1969, Early diagenetic cementation of Recent sands, Fraser River delta, British Columbia: Sedimentology, v. 12, p. 27-46.

Illing, L. V. A., J. Wells, and J. C. M. Taylor, 1965, Penecontemporary dolomite in the Persian Gulf, in L. C. Pray and R. C. Murray, eds., Dolomitization and limestone diagenesis, a symposium: Soc. Econ. Paleontologists and Mineralogists Spec. Pub. No. 13, p. 89-111.

Lattman, L. H., 1973, Calcium carbonate cementation of alluvial fans in southern Nevada: Geol. Soc. America Bull., v. 84, p. 3013-3028.

Muravyov, V. I., 1970, Formation of carbonate cement in clastic rocks: Sedimentology, v. 15, p. 139-145.

Nash, A. J., and E. D. Pittman, 1975, Ferro-magnesian calcite cement in sandstones: Jour. Sed. Petrology, v. 45, p. 258-265.

Roberts, H. H., and T. Wheland, II, 1975, Methane-derived carbonate cements in barrier and beach sands of a subtropical delta complex: Geochim. et Cosmochim. Acta, v. 39, p. 1085-1089.

Shearman, D. J., J. Khouri, and S. Taha, 1961, On the replacement of dolomite by calcite in some Mesozoic limestones from the French Jura: Geologists Assoc. London, Proc., v. 72, p. 1-12.

Smith, R. E., 1969, Petrography-porosity relations in carbonate-quartz system. Gatesville Formation (Late Cambrian), Penn.: AAPG Bull., v. 53, p. 261-278.

Waldschmidt, W. A., 1941, Cementing materials in sandstones and their probable influence on the migration and accumulation of oil and gas: AAPG Bull., v. 25, p. 1839-1879.

Clay and Feldspar

Almon, W. R., L. B. Fullerton, and D. K. Davies, 1976, Pore space reduction in Cretaceous sandstones through chemical precipitation of clay minerals: Jour. Sed. Petrology, v. 46, p. 89-96.

Baptist, O. C., and S. A. Sweeney, 1955, Effects of clays on the permeability of reservoir sands to various saline waters, Wyoming: U.S. Bur. Mines Rept. Inv. 5180, 23 p.

Bartow, J. A., and J. D. Sims, 1975, Authigenic kaolinite associated with Tertiary oil-bearing sandstones, California Coast Ranges, U.S.A.: IX Congress International de Sedimentologie, p. 9-12.

Baskin, Yehuda, 1956, A study of authigenic feldspars: Jour. Geology, v. 64, p. 132-155.

Berg, R. R., 1952, Feldspathized sandstone: Jour. Sed. Petrology, v. 22, p. 221-223.

Borst, R. L., and R. Q. Gregg, 1969, Authigenic mineral growth as revealed by the scanning electron microscope: Jour. Sed. Petrology, v. 39, p. 1596-1597.

Brenchley, P. J., 1969, Origin of matrix in Ordovician graywackes, Berwyn Hills, North Wales: Jour. Sed. Petrology, v. 39, p. 1297-1301.

Bucke, D. P., Jr., and C. J. Mankin, 1971, Clay-mineral diagenesis within interlaminated shales and sandstones: Jour. Sed. Petrology, v. 41, p. 971-981.

Burst, J. F., 1969, Diagenesis of Gulf Coast clayey sediments and its possible relation to petroleum migration: AAPG Bull., v. 53, p. 73-93.

Carrigy, M. A., and G. B. Mellon, 1964, Authigenic clay mineral cements in Cretaceous and Tertiary sandstones of Alberta: Jour. Sed. Petrology, v. 34, p. 461-472.

Cecil, C. B., and M. T. Heald, 1971, Experimental investigation of the effects of grain coatings on quartz growth: Jour. Sed. Petrology, v. 41, p. 582-584.

Cummins, W. A., 1962, The graywacke problem: Liverpool Manchester Geol. Jour., v. 3, p. 51-72.

Divis, A. F., and J. McKenzie, 1975, Experimental authigenesis of phyllosilicates from feldspathic sands: Sedimentology, v. 22, p. 147-155.

Dunoyer de Segonzac, G., 1970, The transformation of clay minerals during diagenesis and low-grade metamorphism: A review: Sedimentology, v. 15, p. 281-346.

Foscolas, A. E., and H. Kodama, 1974, Diagenesis of clay minerals from Lower Cretaceous shales of northeastern British Columbia: Clays and Clay Minerals, v. 22, p. 319-335.

Galloway, W. E., 1974, Deposition and diagenetic alteration of sandstone in northeast Pacific arc-related basin: Implications for graywacke genesis: Geol. Soc. America Bull., v. 85, p. 379-390.

Garrels, R. M., and P. Howard, 1959, Reactions of feldspar and mica with water at low temperatures and pressure: 6th National Conf. Clays and Clay Minerals, Proc., Nat. Acad. Sci., Paper 156, p. 68-88.

Gilbert, C. M., 1949, Cementation of some California Tertiary reservoir sands: Jour. Geology, v. 57, p. 1-17.

Glover, J. E., and P. Hoseman, 1970, Optical data on some authigenic feldspars from western Australia: Mineral Mag., v. 37, p. 588-592.

Goldich, S. S., 1934, Authigenic feldspar in sandstones of southeastern Minnesota: Jour. Sed. Petrology, v. 4, p. 89-95.

Gordon, M., Jr., J. I. Tracey, Jr., and M. W. Ellis, 1958, Geology of the Arkansas bauxite region: U.S. Geological Survey Prof. Paper 299, 268 p.

Grim, R. E., 1958, Concept of diagenesis in argillaceous sediments: AAPG Bull., v. 42, p. 246-253.

——— 1968, Clay mineralogy, 2nd ed.: New York, McGraw-Hill, 596 p.

Hawkins, J. W., Jr., and J. T. Whetten, 1969, Graywacke matrix: Hydrothermal reactions with Columbia River sediments: Science, v. 166, p. 868-870.

Heald, M. T., and R. E. Larese, 1973, The significance of the solution of feldspar in porosity development: Jour. Sed. Petrology, v. 43, p. 458-460.

——— 1974, Influence of coatings on quartz cementation: Jour. Sed. Petrology, v. 44, p. 1269-1274.

Hiltabrand, R. R., R. E. Ferrell, and G. K. Billings, 1973, Experimental diagenesis of Gulf Coast argillaceous sediment: AAPG Bull., v. 57, p. 338-348.

Honess, A. P., and C. D. Jeffries, 1940, Authigenic albite from the Lowville Limestone at Bellefonte, Pennsylvania: Jour. Sed. Petrology, v. 10, p. 12-18.

Jonas, E. C., 1975, Crystal chemistry of clay mineral diagenesis: Internat. Assoc. for the Study of Clay, Proc., Mexico City, p. 1-14.

——— and E. F. McBride, 1977, Diagenesis of sandstone and shale: Application to exploration for hydrocarbons: Austin, Texas, Dept. of Geol. Sciences, Univ. of Texas at Austin, Continuing Education Program, Pub. No. 1, 167 p.

Karpova, G. V., 1969, Clay mineral post-sedimentary ranks in terrigenous rocks: Sedimentology, v. 13, p. 5-20.

Kastner, Miriam, 1971, Authigenic feldspars in carbonate rocks: Am. Mineralogist, v. 56, p. 1403-1442.

Keller, W. D., 1963, Diagenesis in clay minerals—A review: 11th National Conf. Clays and Clay Minerals, Proc., p. 136-157.

Kelley, D. R., and P. F. Kerr, 1957, Clay alteration and ore, Temple Mountain, Utah: Geol. Soc. America Bull., v. 68, p. 1101-1116.

Lerbekmo, J. F., 1957, Authigenic montmorillonoid cement in andesitic sandstones of central California: Jour. Sed. Petrology, v. 27, no. 3, p. 298-305.

——— 1961, Porosity reduction in Cretaceous sandstones of Alberta: Alberta Soc. Petroleum Geol. Jour., v. 9, p. 192-199.

Lovell, J. P. B., 1971, Diagenetic origin of graywacke matrix minerals: A discussion: Sedimentology, v. 19, p. 141-143.

Martin, R. F., 1971, Disordered authigenic feldspars of the series $KAlSi_3O_8$- $KBSi_3O_8$ from southern California: Am. Mineralogist, v. 56, p. 281-291.

Middleton, G. V., 1972, Albite of secondary origin in Charny sandstones, Quebec: Jour. Sed. Petrology, v. 42, p. 341-349.

Millot, Georges, 1970, Geology of clays: New York, Springer-Verlag, 429 p.

——— J. Lucas, and R. Wrey, 1963, Research on evolution of clay minerals and argillaceous and siliceous neoformation: 10th National Conf. Clays and Clay Minerals: New York, Pergamon, p. 399-412.

Müller, G., 1967, Diagenesis in argillaceous sediments, in G. Larsen and G. V. Chilingar, eds., Devel. in Sedimentology, Volume 8: Diagenesis in sediments, New York, Elsevier, p. 127-177.

Odom, I. E., 1975, Feldspar-grain size relations in Cambrian arenites, upper Mississippi Valley: Jour. Sed. Petrology, v. 45, p. 636-650.

Perry, E., and J. Hower, 1970, Burial diagenesis in Gulf Coast pelitic sediments: Clays and Clay Minerals, v. 18, p. 165-177.

——— 1972, Late stage dehydration in deeply buried pelitic sediments: AAPG Bull., v. 56, p. 2013-2021.

Pittman, E. D., and D. N. Lumsden, 1968, Relationship between chlorite coatings on quartz grains and porosity, Spiro sand, Oklahoma: Jour. Sed. Petrology, v. 38, p. 668-670.

Sarkisyan, S. G., 1971, Application of the scanning electron microscope in the investigation of oil and gas reservoir rocks: Jour. Sed. Petrology, v. 41, p. 289-292.

——— 1972, Origin of authigenic clay minerals and their significance in petroleum geology: Sed. Geology, v. 7, p. 1-22.

Shelton, J. W., 1964, Authigenic kaolinite in sandstone: Jour. Sed. Petrology, v. 34, p. 102-111.

Sibley, D. F., 1978, K-feldspar cement in the Jacobsville Sandstone: Jour. Sed. Petrology, v. 48, p. 983-986.

Sims, J. D., 1970, Authigenic kaolinite in sand of the Wilcox Formation, Jackson Purchase region, Kentucky: U.S. Geol. Survey Prof. Paper 700-B, p. 1327-1332.

Stablein, N. K., III, and E. C. Dapples, 1977, Feldspars of the Tunnel City Group (Cambrian), western Wisconsin: Jour. Sed. Petrology, v. 47, p. 1512-1538.

Stadler, P. J., 1973, Influence of crystallographic habit and aggregate structure of authigenic clay minerals on sandstone permeability: Geol. en Mijnbouw, v. 52, p. 217-220.

Swett, Keene, 1968, Authigenic feldspars and cherts resulting from dolomitization of illitic limestones: A hypothesis: Jour. Sed. Petrology, v. 38, p. 128-135.

Tester, A. C., and G. I. Atwater, 1934, The occurrence of authigenic feldspars in sediments: Jour. Sed. Petrology, v. 4, p. 23-31.

Triplehorn, D. M., 1970, Clay mineral diagenesis in Atoka (Pennsylvanian) sandstones, Crawford County, Arkansas: Jour. Sed. Petrology, v. 40, p. 838-847.

Weaver, C. E., 1960, Possible uses of clay minerals in search for oil: AAPG Bull., v. 44, p. 1505-1518.

—— and K. C. Beck, 1971, Clay water diagenesis during burial: how mud becomes gneiss: Geol. Soc. America Spec. Paper 134, 96 p.

Webb, J. E., 1974, Relation of oil migration to secondary clay cementation, Cretaceous sandstones, Wyoming: AAPG Bull., v. 58, p. 2245-2249.

Whetten, J. T., and J. W. Hawkins, Jr., 1970, Diagenetic origin of graywacke matrix minerals: Sedimentology, v. 15, p. 347-361.

—— and R. R. Hiltabrand, 1972, Formation of authigenic clay in detrital sand: AAPG Bull., v. 56, p. 662.

Wilson, M. D., and E. D. Pitman, 1977, Authigenic clays in sandstones: recognition and influence on reservoir properties and paleoenvironmental analysis: Jour. Sed. Petrology, v. 47, p. 3-31.

Zeolites

Boles, J. R., and D. S. Coombs, 1977, Zeolite facies alteration of sandstones in Southland Syncline, New Zealand: Am. Jour. Sci., v. 277, no. 8, p. 982-1012.

Deffeyes, K. S., 1959, Zeolites in sedimentary rocks: Jour. Sed. Petrology, v. 29, p. 602-609.

Gilbert, C. M., and M. G. McAndrews, 1948, Authigenic heulandite in sandstone, Santa Cruz County, California: Jour. Sed. Petrology, v. 18, p. 91-99.

Hay, R. L., 1963, Stratigraphy and zeolitic diagenesis of the John Day formation of Oregon: Univ. of Calif. Pub. in Earth Sciences, v. 42, no. 5, p. 199-262.

—— 1966, Zeolites and zeolitic reactions in sedimentary rocks: Geol. Soc. America Spec. Paper 85, 130 p.

Heath, G. R., 1969, Mineralogy of Cenozoic deep-sea sediments from the equatorial Pacific Ocean: Geol. Soc. America Bull., v. 80, p. 1997-2018.

Kaley, M. E., and R. F. Hanson, 1955, Laumontite and leonhardite in Miocene sandstone from a well in San Joaquin Valley, California: Am. Mineralogist, v. 40, p. 923-925.

Lerbekmo, J. F., 1957, Authigenic montmorillonoid cement in adesitic sandstone: Jour. Sed. Petrology, v. 27, p. 298-305.

Munson, R. A., and R. A. Sheppard, 1974, Natural zeolites: their properties, occurrences, and uses: Minerals Sci. Engineering, v. 6, p. 19-34.

Murata, K. J., and K. R. Whiteley, 1973, Zeolites in the Miocene Briones Sandstone and related formations of the central Coast Ranges, California: U.S. Geol. Survey Jour. Research, v. 1, p. 255-265.

Sheppard, R. A., and A. J. Gude, 3rd, 1968, Distribution and genesis of authigenic silicate minerals in tuffs of Pleistocene Lake Tecopa, Inyo County, California: U.S. Geol. Survey Prof. Paper 597, 38 p.

—— 1969, Diagenesis of tuffs in the Barstow Formation, Mud Hills, Sand Bernardino County, California: U.S. Geol. Survey Prof. Paper 634, 35 p.

—— 1973, Zeolites and associated authigenic silicate minerals in tuffaceous rocks of the Big Sandy Formation, Mojave County, Arizona: U.S. Geol. Survey Prof. Paper 830, 36 p.

Van Houten, R. B., 1962, Cyclic sedimentation and the origin of analcime-rich Upper Triassic Lockatong, west-central New Jersey and adjacent Pennsylvania: Am. Jour. Sci., v. 260, p. 561-576.

Phosphates

Cook, P. J., 1970, Repeated diagenetic calcitization, phosphatization, and silicification in the Phosphoria Formation: Geol. Soc. America Bull., v. 81, p. 2107-2116.

Iron and Manganese

Adeleye, D. R., 1973, Origin of ironstones, an example from the Middle Niger valley, Nigeria: Jour. Sed. Petrology, v. 43, p. 709-727.

Berner, R. A., 1970, Sedimentary pyrite formation: Am. Jour. Sci., v. 268, p. 1-24.

Bonatti, E., and Y. R. Nayudu, 1965, The origin of manganese nodules on the ocean floor: Am. Jour. Sci., v. 263, p. 17-39.

Kimberley, M. M., 1974, Origin of iron ore by diagenetic replacement of calcareous oolite: Nature, v. 250, p. 319-320.

McBride, E. F., 1974, Significance of color in red, green, purple, olive, brown, and gray beds of Difunta Group, northeastern Mexico: Jour. Sed. Petrology, v. 44, p. 760-773.

McGeary, D. F. R., and J. E. Damuth, 1973, Postglacial iron-rich crusts in hemipelagic deep-sea sediment: Geol. Soc. America Bull., v. 84, p. 1201-1212.

Pequegnat, W. E., W. R. Bryant, A. D. Fredericks, T. R. McKee, and R. Spalding, 1972, Deep-sea ironstone deposits in the Gulf of Mexico: Jour. Sed. Petrology, v. 42, p. 700-710.

Price, N. B., 1967, Some geochemical observations on manganese-iron oxide nodules from different depth environments: Marine Geology, v. 5, p. 511-538.

Price, W. A., 1962, Stages of oxidation coloration in dune and barrier sands with age: Geol. Soc. America Bull., v. 73, p. 1281-1284.

Schluger, P. R., and H. E. Roberson, 1975, Mineralogy and chemistry of the Patapsco Formation, Maryland, related to the ground-water geochemistry and flow system: A contribution to the origin of red beds: Geol. Soc. America Bull., v. 86, p. 153-158.

Sugden, W., 1966, Pyrite staining of pellety debris in carbonate sediments from the Middle East and elsewhere: Geol. Mag., v. 103, p. 250-256.

Van Houten, F. B., 1963, Origin of red beds—some unsolved problems, in A. E. M. Nairn, ed., Problems in paleoclimatology: New York, Interscience Pub., p. 647-659.

—— 1968, Iron oxides in red beds: Geol. Soc. America Bull., v. 79, p. 399-416.

—— 1973, Origin of red beds, a review—1961-1972: Ann. Review Earth and Planet. Sci., v. 1, p. 39-61.

Walker, T. R., 1967a, Formation of red beds in modern and ancient deserts: Geol. Soc. America Bull., v. 78, p. 353-368.

—— 1967b, Color of Recent sediments in tropical Mexico: A contribution to the origin of red beds: Geol. Soc. America Bull., v. 78, p. 917-920.

—— 1974, Formation of red beds in moist tropical climates: A hypothesis: Geol. Soc. America Bull., v. 85, p. 633-638.

—— 1976, Diagenetic origin of continental red beds, in H. Falke, ed., The continental Permian in central, west, and south Europe: Dordrecht-Holland, D. Riedel Pub. Co., p. 240-282.

—— and R. M. Honea, 1969, Iron content of modern deposits in the Sonoran desert: A contribution to the origin of red beds: Geol. Soc. America Bull., v. 80, p. 535-544.

—— Brian Waugh, and A. J. Grove, 1978; Diagenesis in first-cycle desert alluvium of Cenozoic age, southwestern United States and northwestern Mexico: Geol. Soc. America Bull., v. 89, p. 19-32.

Replacement
or
Displacement
Fabrics

Upper Cretaceous Monte Antola Fm.
Italy

Authigenic pyrite crystals. Pyrite commonly forms as a replacement of organic matter or in close proximity to zones of concentrated organic material. Opaque in reflected light, pyrite is most readily identified using reflected light because of its characteristic yellow-gold metallic appearance. This example is probably a replacement of detrital plant fragments.

0.10 mm

Upper Triassic Brunswick Fm. ★
New Jersey

Authigenic pyrite crystals which have formed as a replacement or displacement in shale. Although detrital and synsedimentary pyrite occur, most pyrite in sedimentary rocks is of diagenetic origin. Crystals are generally cubes, pyritohedrons, or octahedrons, and they form most commonly under reducing conditions.

0.27 mm

Upper Triassic Brunswick Fm. ★
New Jersey

A view similar to the one above but using reflected light. Cubic pyrite crystals of authigenic origin are scattered through an illitic- and chloritic- clay matrix in this lacustrine sediment. Pyrite is readily identifiable here because of its characteristic yellowish, metallic appearance and its cubic crystal form.

RL   0.22 mm

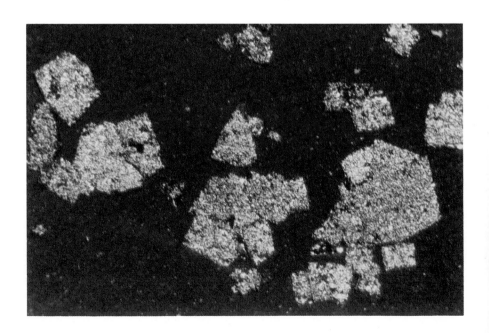

Upper Cretaceous Atco Fm.
 (Austin Group)
Texas

An SEM view of authigenic pyrite crystals in a limestone. As with most pyrite, this example shows euhedral crystals which partly interlock and which have formed by replacement.

SEM   35 μm

Upper Cretaceous Chalk
British North Sea 378 m (1,240 ft)

SEM view of framboidal pyrite. Framboids are small clusters of minute, interlocking pyrite crystals with smooth, almost perfectly spherical exterior surfaces. These can form singly or in groups as authigenic replacements in limestones and other sedimentary rocks.

SLM   3 μm

Devonian Cairn Fm.
Canada (Alberta)

Authigenic quartz replacing a carbonate rock. Evidence for the authigenic origin of these quartz grains includes the euhedral terminations, the large number of carbonate inclusions, and the cross-cutting textural relations. Such replacement crystals commonly (though not always) form as very large overgrowths of detrital quartz silt or sand grains.

XN   0.27 mm

Devonian Cairn Fm. *
Canada (Alberta)

Authigenic quartz crystals in a limestone. Numerous crystals with euhedral, bipyramidal outlines can be seen. In some cases, a nucleus of detrital quartz silt can be seen. Note presence of abundant carbonate inclusions within these crystals of replacement origin.

XN   0.27 mm

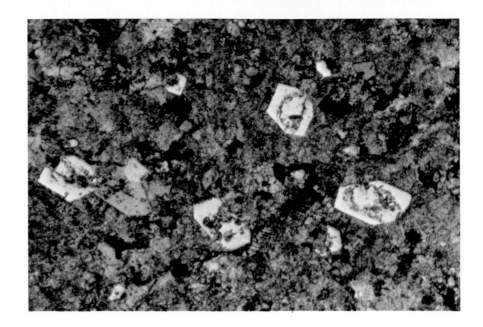

Cambrian Conococheague Ls.
Virginia

Authigenic quartz in a limestone. Although the quartz grain in the center of the photo appears to be a replacement product, its position within a small, indistinct vein and its lack of inclusions of carbonate sediment argue for a pore filling origin within an originally unfilled fracture. Careful petrographic observation commonly is necessary to distinguish replacement versus void-filling textures.

0.38 mm

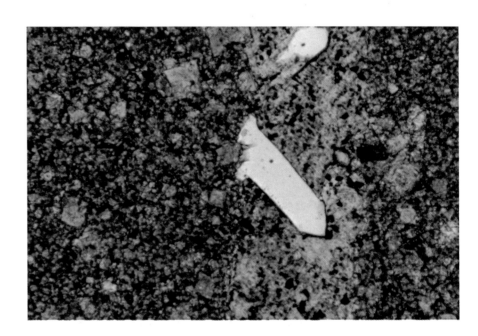

Oligocene Frio Fm.
Texas 3,068 m (10,066 ft)

Authigenic quartz crystals in a sandstone. In SEM, authigenic quartz is recognized primarily by its euhedral, bipyramidal outline. Although textural relations are commonly difficult to decipher in SEM, this example appears to show quartz forming as a void filling rather than as a replacement. Note orientation of crystals and increase of crystal size toward viewer (cavity center).

SEM   3.3 μm

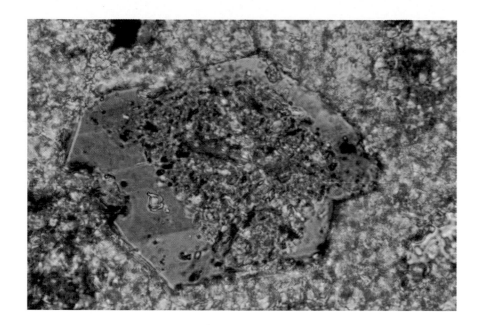

Upper Cretaceous Monte Antola Fm.
Italy

Authigenic-feldspar replacement of lime-stone. Authigenic feldspars in carbonate rocks are most commonly (as here) of albite composition although microcline also occurs. Detrital nuclei are commonly present and abundant carbonate inclusions serve to confirm an authigenic origin.

XN   0.06 mm

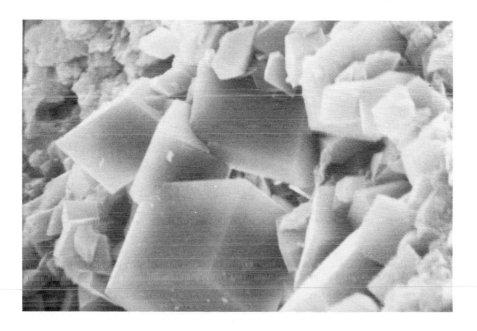

Pliocene Big Sandy Fm.
Arizona

Authigenic feldspar crystals in SEM. These are potassium feldspars which have formed as a replacement of analcime (a zeolite) in tuffaceous lacustrine rocks. The crystals are euhedral to subhedral, partly inter-grown, and 4 to 8 μm in size. Photo by A. J. Gude, III.

SEM   3 μm

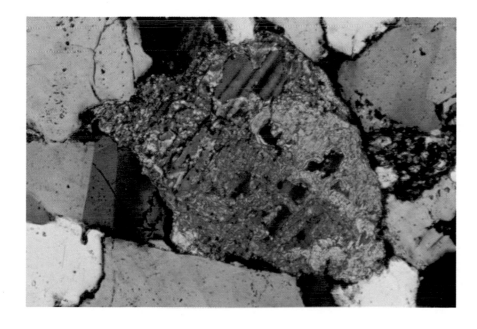

Pennsylvanian-Permian Sangre
 de Cristo Fm.
New Mexico

Calcite replacement of a detrital plagioclase feldspar. This is a fairly common type of replacement; subsequent dissolution of the replacement calcite can account for much of the secondary porosity associated with feldspars. Note relicts of twinned feldspar still preserved within the calcite crystals. Calcite replacement is most common in calcic plagioclases and may selectively replace certain zones.

XN   0.08 mm

Upper Cretaceous Star Point Ss.
Utah

Dolomite in a chert clast. Determination of a paragenetic sequence in this case is quite difficult. The chert (white) is probably a replacement of limestone. The dolomite (brownish rhombs) was either an early replacement of the limestone not replaced by the chert, or the dolomite was formed as a late diagenetic replacement of the chert. Such ambiguities are commonly not resolvable in single clasts or samples.

0.27 mm

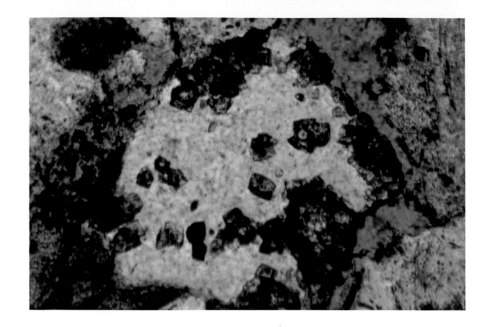

Cretaceous Schisti Galestrini *
Italy   •

Chert replacement of dolomite. The photo shows a well-developed rhombic fabric with some internal zonation. The fabric is clearly that of dolomite replacement of limestone; the entire rock is, however, composed of chert. Only through petrographic analysis could one establish that this rock has progressed from a limestone to a dolomite, and, finally, to a chert.

0.08 mm

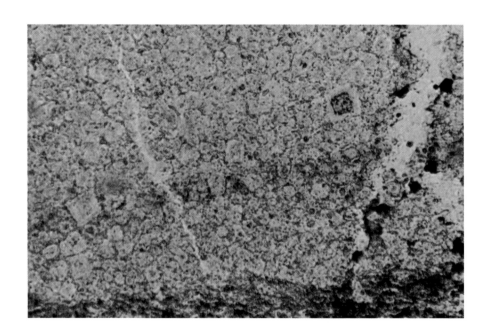

Cretaceous Schisti Galestrini *
Italy   .

Same view as previous photo but with crossed polarizers. Under cross-polarized light one can see that the rock consists of chert and megaquartz. Some zones of dolomite are still preserved, as in the large rhomb in the upper right. Clays are a significant component in some areas such as along the bottom edge of the photo.

XN   0.08 mm

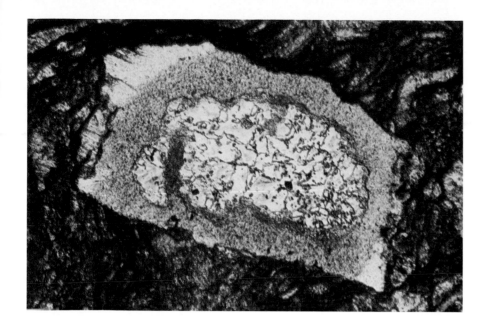

Silurian Tonoloway-Keyser Ls.
Pennsylvania

Chalcedonic quartz replacement of a calcitic fossil. The grain being replaced is a crinoid columnal with small syntaxial overgrowths at both ends of the long axis. Chalcedony has selectively replaced the central portion of this fossil. Selectivity in replacement is a very common phenomenon with individual fossil types, compositional zones, or other features being preferentially altered.

XN   0.15 mm

Upper Jurassic Morrison Fm.,
  Salt Wash Mbr.
Colorado

Although quartz is a very stable mineral, it too is subject to replacement by other minerals under certain conditions. Here, detrital quartz is partially replaced by a uranium-bearing mineral (uraninite or coffinite) which consists of radiating bundles of opaque, bladed crystals. Brown interstitial material between detrital grains is clay.

0.06 mm

Pennsylvanian Strawn Gp. "Gray Ss."
Texas 1,492 m (4,896 ft)

Vermicular kaolinite within a detrital clay clast. Although this may appear to be a direct replacement of one clay mineral by another, it is more likely (because of the well-developed vermicular texture and inter-crystalline porosity) that the kaolinite is an authigenic filling of a secondary pore created by partial leaching of the original clay clast. Photo by S. P. Dutton.

0.09 mm

Deformation Fabrics

Cambrian Conococheague Fm.
Virginia

Fracturing in a dolomitic chert. Matrix is a dark chert (probably a replacement of limestone) which has been partly dolomitized. In this case, fracturing clearly postdates dolomitization as evidenced by the displacement of the two halves of the dolomite crystal. The fracture has been filled with equant megaquartz. The existence and relative timing of fractures commonly can be determined petrographically.

XN   0.08 mm

Jurassic-Cretaceous shale
Alaska

A fractured limestone in which the fracture has been filled with unusual, elongate calcite spar crystals mixed with authigenic quartz (gray). Such fractures generally form rather late in the diagenetic history of sediments for fracturing implies a considerable measure of prior lithification.

XN   0.30 mm

Eocene Crescent Fm. ★
Washington

Calcite-filled fractures in a dark shale. These fractures may have formed at an intermediate stage in the diagenetic history of this rock for subsequent compaction has led to shearing of the veins and intense twinning of the calcite crystals.

XN   0.10 mm

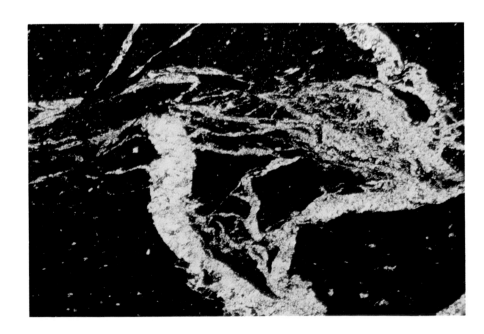

Devonian Cairn Fm. *
Canada (Alberta)

A quartz-filled fracture in limestone. The quartz shows a typical void filling morphology with equant crystals increasing in size from the fracture walls to the center of the void. This indicates that fracture porosity once was present in this sample.

XN   0.07 mm

Lower Cambrian St. Roch Fm.
Canada (Quebec)

Quartz-filled veins in a tightly cemented sandstone. The intersecting fractures are filled with coarse, elongate crystals of megaquartz which entirely span the width of the vein. The quartz crystals are rich in bubble inclusions, a common feature in hydrothermal veins.

XN   0.27 mm

Oligocene Tongriano Cgl.
Italy

A fracture filled with megaquartz and calcite. Quartz lines the fracture walls or completely fills parts of the vein. Calcite has apparently grown after the quartz or during late stages of quartz precipitation and fills remnant void space. The intense twinning of the calcite crystals indicates that vein filling preceeded the complete termination of deformation.

XN   0.27 mm

Cambrian Sillery Group
Canada (Quebec)

Individual detrital grains can also show evidence of structural deformation and compaction. This example illustrates relatively soft shale clasts which were compacted during burial, and which flowed around adjacent quartz grains. Such compaction fabrics are common in rocks rich in clay or shale clasts, in part because of the softness of such grains, but also because the detrital clays may inhibit overgrowth cementation.

0.27 mm

Cambrian(?) Unicoi Fm.
Virginia

An extreme example of the deformation of detrital shale clasts. Clasts here have flowed between adjacent harder fragments (mainly quartz) and, in places, virtually simulate a dispersed clay matrix. Compaction of ductile grains can be an important factor in porosity loss in sandstones, especially those with abundant glauconite grains or shale clasts.

XN   0.10 mm

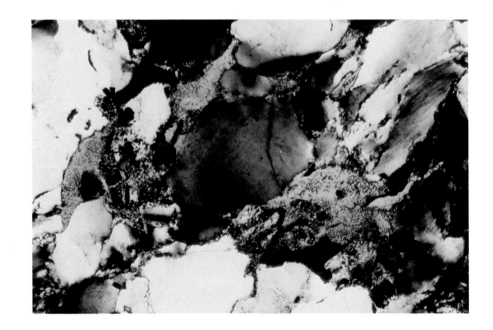

Pennsylvanian-Permian Sangre
  de Cristo Fm. *
New Mexico

Individual detrital minerals which are especially flexible or brittle may also act as sensitive indicators of deformation. In this example, micas (muscovite) have been bent and fractured by adjacent, harder grains. Some mutual interpenetration of quartz and feldspar grains is also evident. Such compactional or deformational alteration can significantly reduce porosity without cementation.

XN   0.15 mm

Upper Cretaceous Monte Antola Fm.
Italy

Grain deformation in a sandy limestone.
The muscovite flake in the center was
deformed during compaction of relatively
uncemented material. This is one of several
lines of evidence indicating that this
material was not cemented until relatively
late in its diagenetic history. In such cases
compaction may account for most of the
overall porosity reduction.

XN   0.15 mm

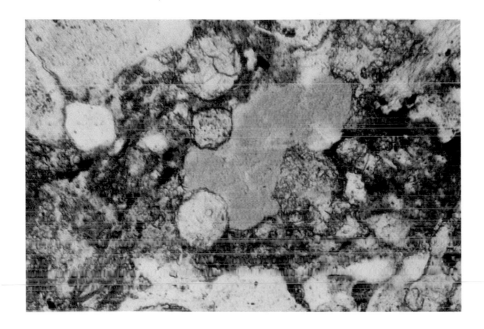

Upper Cretaceous Monte Antola Fm.
Italy

Compactional deformation of glauconite.
Because glauconite pellets are quite soft,
they are especially susceptible to compaction. Here, calcite grains (replaced circular
sections of sponge spicules) deeply embay
the glauconite.

0.06 mm

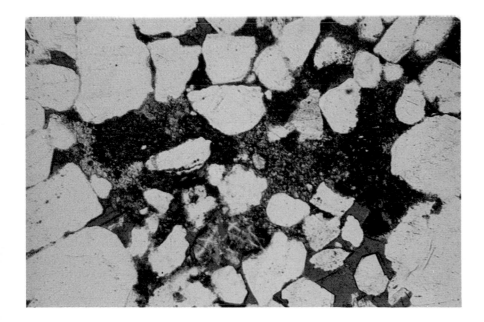

Lower Cretaceous sandstone
Canada (N.W.T.) 2,977 m (9,833 ft)

Compaction of eogenetic siderite nodules
and pellets. Deformation of the siderite
grains preceeded the establishment of a
stable cement framework and extensively
reduced porosity by flowing around
adjacent quartz grains. Subsequent formation of thin quartz overgrowths and extensive poikilotopic calcite cement (stained
red) completed porosity obliteration.
Photo by V. Schmidt.

0.22 mm

Upper Triassic New Haven Arkose
Connecticut

Brittle grains, especially those with well
developed fracture or cleavage directions,
are likely to show deformation effects
readily. In this example, a plagioclase
feldspar has been fractured parallel with
its twin planes through compaction.
Adjacent quartz grains have been forced
into the plagioclase along one margin.

XN   0.15 mm

Cambrian Hickory Ss. Mbr.
 of Riley Fm. *
Texas

In zones of intense deformation, such as
along fault traces, even minerals such as
quartz may be deformed or fractured.
Here, two adjacent quartz grains have
developed a series of fractures which
radiate out from the original point of
contact between the grains.

XN   0.10 mm

Cambrian Hickory Ss. Mbr.
 of Riley Fm.
Texas

This example, from the same rock as the
previous photo, shows the widespread
fracturing of quartz grains in this sample.
The deformation and fracturing of these
grains is clearly post-depositional as evidenced by the continuity of fracturing
from one grain to the next, the abundance
of fractured grains, and the fact that such
intensely fractured grains would not have
survived transport.

XN   0.27 mm

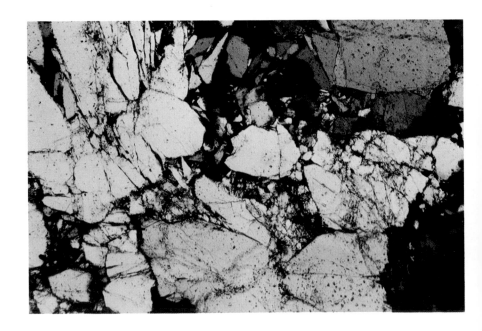

Upper Cambrian Gatesburg Fm.
Pennsylvania

A calcitic ooid which has been subjected to intense compactional deformation. Note shearing of concentric oolitic coatings from the nucleus of the ooid. Such deformation indicates that the rock was not firmly cemented until rather late in its diagenetic history.

GP   0.30 mm

Lower Cretaceous Patula Arkose
Mexico

A detrital quartz grain with warped, sub-parallel lines of very small bubble inclusions. These have been called Boehm lamellae and are the product of intense strain in quartz. When found in a majority of the quartz grains in a rock, they can be important evidence of *in situ* deformation.

XN   0.10 mm

Devonian Oriskany Ss.
Virginia

A probable grain-compaction and pressure-solution texture. Quartz grains are well rounded, show no euhedral overgrowths, and are coated with thin clay films. Most grains have irregular line contacts with adjacent minerals rather than the usual point-to-point contacts found in undeformed sediments. Many of the quartz grains also have strain shadows and Boehm lamellae. Thus, it is likely that this sample underwent significant deformation.

XN   0.15 mm

Lower Devonian Becraft Ls.
New York

Intergranular pressure solution is most common in limestones. In this example, a large brachiopod shell (upper part of photo) and numerous syntaxially cemented echinoderm fragments have mutually dissolved and interpenetrated each other along a high amplitude stylolite surface. Relatively little insoluble material has been concentrated along the stylolite.

0.38 mm

Jurassic Morrison Fm., Salt Wash Mbr.
Colorado

A stylolitic solution surface in a sandstone. The dark zone which passes through the center of the photo is the stylolite interval in which insoluble material, such as clays or organic matter, has been concentrated. Note truncation of rounded quartz grains along stylolite. Such stylolites may provide a source of silica for subsurface quartz cementation.

0.27 mm

Lower Ordovician Phycoden Schichten
Germany

Solution seams (low amplitude stylolites) in a siltstone. In this case, solution seams are most common adjacent to (or between) cemented burrows filled with coarser material (lighter colored areas at top and bottom of photo). Compactional drape and pressure solution took place most effectively around such harder zones. Note multiple seams of concentrated insoluble residue.

0.22 mm

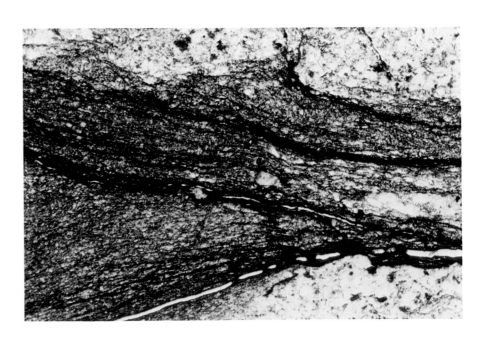

Selected Compaction and Fracturing Bibliography

Athy, L. F., 1930, Density, porosity and compaction of sedimentary rocks: AAPG Bull., v. 14, p. 1-24.

Baldwin, Brewster, 1971, Ways of deciphering compacted sediments: Jour. Sed. Petrology, v. 41, p. 293-301.

Brown, P. R., 1969, Compaction of fine-grained terrigenous and carbonate sediments—A review: Canadian Petrol. Geol. Bull., v. 17, p. 486-495.

Chilingarian, G. V., and H. H. Rieke, III, 1968, Data on consolidation of fine-grained sediment: Jour. Sed. Petrology, v. 38, p. 811-816.

—— and K. H. Wolf, eds., 1975, Compaction of coarse-grained sediments, vol. 1: New York, Elsevier, Developments in Sedimentology, v. 18a, 552 p.

—— —— eds., 1976, Compaction of coarse-grained sediments, vol. 2: New York, Elsevier, Developments in Sedimentology, v. 18b, 808 p.

Deelman, J. C., 1975a, An experimental approach to lithification textures: Nature, v. 257, no. 5529, p. 782-783.

—— 1975b, Pressure solution or indentation: Geology, v. 3, p. 23-24.

Gipson, M., Jr., 1966, A study of the relations of depth, porosity and clay mineral orientation in Pennsylvanian shales: Jour. Sed. Petrology, v. 36, p. 888-903.

Heald, M. T., 1955, Stylolites in sandstones: Jour. Geology, v. 63, p. 101-114.

—— 1959, Significance of stylolites in permeable sandstones: Jour. Sed. Petrology, v. 29, p. 251-253.

Lerbekmo, J. F., and R. L. Platt, 1962, Promotion of pressure-solution of silica in sandstones: Jour. Sed. Petrology, v. 32, p. 514-519.

Maxwell, J. C., 1960, Experiments on compaction and cementation of sand, in D. Griggs and J. Handin, eds., Rock deformation: Geol. Soc. America Mem. 79, p. 105-132.

—— 1964, Influence of depth, temperature, and geologic age on porosity of quartzose sandstone: AAPG Bull., v. 48, no. 5, p. 697.

Meade, R. H., 1966, Factors influencing the early stages of the compaction of clays and sands—a review: Jour. Sed. Petrology, v. 36, p. 1085-1101.

Renton, J. J., M. T. Heald, and C. B. Cecil, 1969, Experimental investigation of pressure solution of quartz: Jour. Sed. Petrology, v. 39, p. 1107-1117.

Rieke, H. H., III, and G. V. Chilingarian, 1974, Compaction of argillaceous sediments; Devel. in Sedimentology, Volume 16: New York, Elsevier, 424 p.

Rittenhouse, Gordon, 1971, Mechanical compaction of sands containing different percentages of ductile grains: A theoretical approach: AAPG Bull., v. 55, p. 92-96.

Sibley, D. F., 1975, Extension microfractures in the Tuscarora orthoquartzite: Evidence from luminescence petrography: Nuclide Spectra, v. 8, no. 1, April 1975.

—— and H. Blatt, 1976, Intergranular pressure solution and cementation of the Tuscarora orthoquartzite: Jour. Sed. Petrology, v. 46, p. 881-896.

Sippel, R. F., 1968, Sandstone petrology, evidence from luminescence petrography: Jour. Sed. Petrology, v. 38, p. 530-554.

Sloss, L. L., and D. E. Feray, 1948, Microstylolites in sandstone: Jour. Sed. Petrology, v. 18, p. 3-13.

Thompson, A., 1959, Pressure solution and porosity, in H. A. Ireland, ed., Silica in sediments: Soc. Econ. Paleontologists and Mineralogists Spec. Pub. No. 7, p. 92-110.

Trurnit, P., 1968, Pressure solution phenomena in detrital rocks: Sedimentary Geology, v. 2, p. 89-114.

Weller, J. M., 1959, Compaction of sediments: AAPG Bull., v. 43, p. 273-310.

Weyl, P. K., 1959, Pressure solution and the force of crystallization—A phenomenological theory: Jour. Geophys. Research, v. 64, p. 2001-2025.

Porosity

Porosity Classification
(definitions, criteria)

Sandstone Porosity Classification
(modified from Schmidt, McDonald, and Platt, 1977 and
Choquette and Pray, 1970)

PRIMARY

Interparticle—very common.

Intraparticle—rare, but possible within rock fragments, fossils, and other detrital grains.

Intercrystal—rare? Remnant primary porosity can be significant within clay cements.

SECONDARY

Dissolution of detrital grains—common, especially removal of feldspars, carbonate rock fragments or fossils, or detrital sulphates.

Dissolution of authigenic cements—very common removal of calcite, dolomite, and siderite; significant removal of gypsum and/or anhydrite.

Dissolution of authigenic replacement minerals—common removal of carbonate or sulphate minerals.

Shrinkage—minor; can be significant in glauconitic sediments.

Fracturing—minor except locally.

Time of formation of diagenetic features
(after Choquette and Pray, 1970)

Eogenetic—occurring during the time interval between final deposition and burial of the newly deposited sediment or rock below the depth of significant influence by processes that either operate from the surface or depend for their effectiveness on proximity to the surface.

Mesogenetic—occurring during the time interval in which rocks or sediments are buried at depth below the major influence of processes directly operating from or closely related to the surface.

Telogenetic—occurring during the time interval during which long buried sediments or rocks are influenced significantly by processes associated with the formation of an unconformity.

Genetic modifiers
(after Choquette and Pray, 1970)

Preserved
Enlarged
Reduced
Filled

Petrographic criteria for secondary porosity.
(from: Schmidt, McDonald, and Platt, 1977)

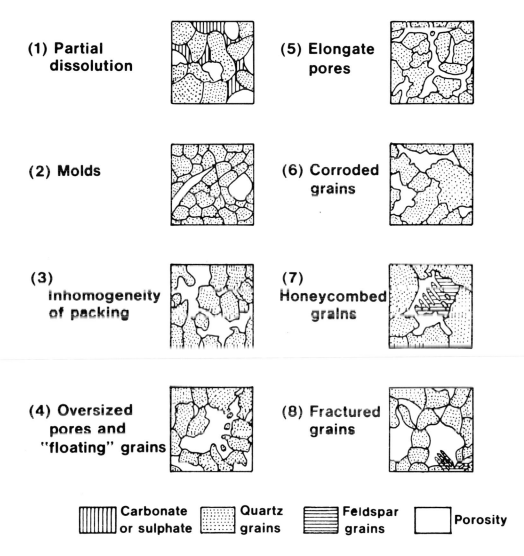

(1) Partial dissolution

(2) Molds

(3) Inhomogeneity of packing

(4) Oversized pores and "floating" grains

(5) Elongate pores

(6) Corroded grains

(7) Honeycombed grains

(8) Fractured grains

Carbonate or sulphate   Quartz grains   Feldspar grains   Porosity

172

Upper Cretaceous Upper Logan Canyon Ss.
Canada (Scotian Shelf) 1,548 m (5,080 ft)

Porosity types in sandstones can be well
classified using the scheme of Schmidt,
McDonald, and Platt (1977) shown on the
preceeding pages, coupled with modifiers
used by Choquette and Pray (1970). This
sample shows high preserved primary
interparticle porosity (blue stain). Detrital
grains are mainly quartz.   Photo by D. A.
McDonald.

0.20 mm

Upper Cretaceous Kogosukruk Tongue of
Prince Creek Formation *
Alaska

Partial cementational loss of porosity—
reduced interparticle porosity. Cementation has been accomplished mainly by
quartz and feldspar overgrowth formation.

0.15 mm

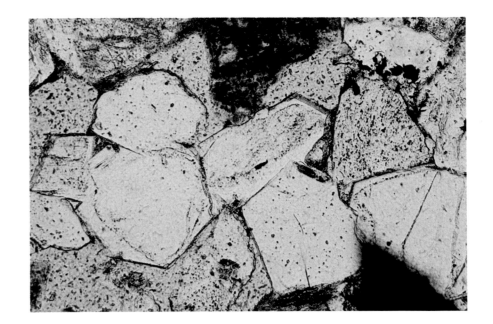

Lower Cretaceous sandstone
Canada

Complete destruction of primary porosity
by cementation—filled interparticle porosity. Cementation is primarily by quartz
overgrowth, easily detected in this case
because of the clear "dust rims" between
detrital grain cores and the authigenic
overgrowths. Photo by V. Schmidt.

PXN   0.20 mm

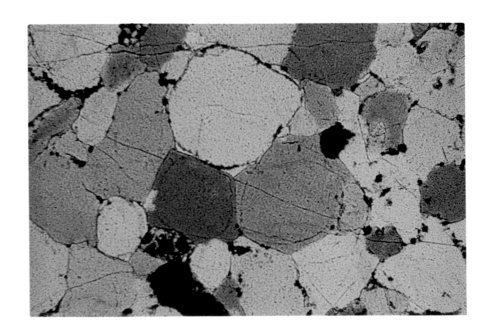

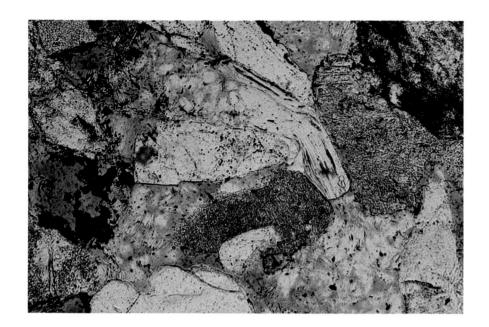

Triassic Dockum Gp.
Texas

Largely filled interparticle porosity with extensive remnant intercrystal porosity. Cementation is primarily by kaolinite and, although overall permeability is greatly reduced, significant porosity (shown in blue) remains between kaolinite "worms." The use of deeply stained impregnating media is essential when studying porosity, especially for the recognition of intercrystalline porosity and to distinguish grain plucking during section grinding from original porosity.

0.05 mm

Oligocene Frio Fm.
Texas 4,825 m (15,829 ft)

Complex porosity in a sandstone. Reduced primary interparticle and intercrystalline porosity as well as fabric selective dissolution porosity (secondary removal of detrital grains) are all present. The secondary porosity has resulted from incipient stages of dissolution of plagioclase feldspar. Note quartz overgrowth cementation. Photo by R. G. Loucks.

0.10 mm

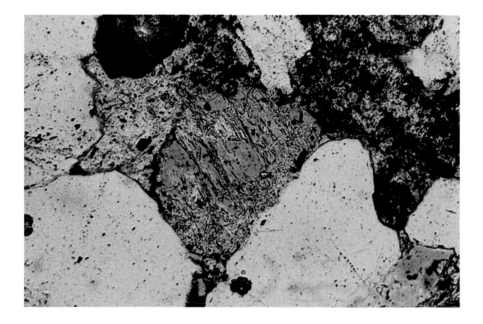

Triassic Dockum Gp.
Texas

Partial dissolution of detrital grains has yielded leached (secondary) porosity. Compaction and cementation have virtually completely obliterated primary porosity but significant amounts of secondary porosity have been generated by the dissolution of feldspars and, possibly, other unstable minerals. Such secondary porosity zones must be interconnected by other porosity types (or be extremely abundant) to act as "effective" pore space.

0.10 mm

Miocene 'Hayner Ranch Fm.'
New Mexico

Dissolution porosity seen in SEM. Secondary leaching of plagioclase feldspars, calcitic replacements of former feldspars, calcite cements, unstable detrital heavy minerals, evaporites, and other minerals is a common feature in sandstones. Note crystallographic control of dissolution in this example. Photo by C. W. Keighin, courtesy of T. R. Walker.

SEM   10 μm

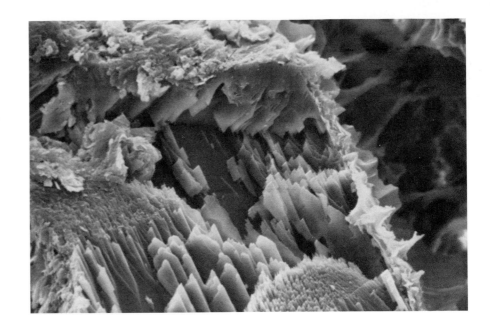

Oligocene Frio Fm. *
Texas 4,837 m (15,870 ft)

An incompletely leached plagioclase feldspar with authigenic feldspar overgrowths partially filling the secondary pore space produced by leaching. Quartz overgrowths have partially filled primary intergranular pore space. Such complex porosity patterns can be deciphered only through petrographic studies. Photo by R. G. Loucks.

0.11 mm

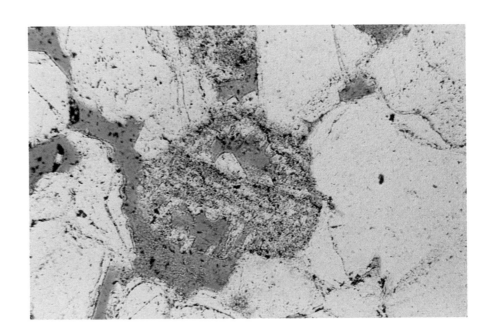

Upper Cretaceous Frontier Fm.
Wyoming ca. 610 m (2,000 ft)

A partially leached plagioclase feldspar with subsequently formed authigenic feldspar overgrowths. Complex porosity development and reduction histories can be unravelled using the SEM in combination with standard petrographic studies. Photo by E. D. Pittman.

SEM   20 μm

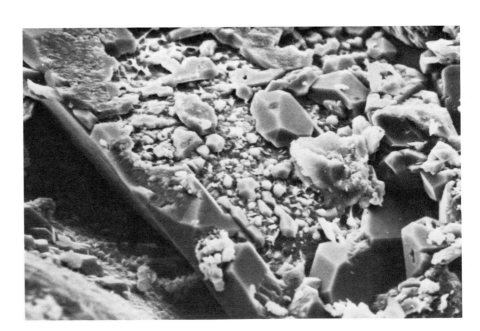

Oligocene Frio Fm.
Texas 920 m (3,017 ft)

Leaching of feldspars (plagioclase) from a volcanic rock fragment. Subsurface dissolution of unstable grains can commonly be quite selective. In some cases, individual compositional zones are removed from feldspars or individual feldspar types are removed from complex rock fragments.

0.07 mm

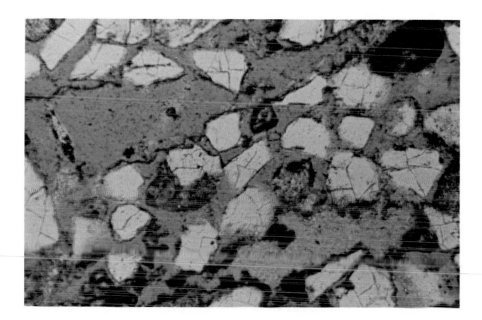

Oligocene Frio Fm. ★
Texas 3,069 m (10,068 ft)

Abundant secondary porosity coupled with partially preserved primary porosity. The leaching of grains is recognizable, in this example, through the presence of molds of former particles, outlined by rims of clay and siderite. The pores left behind through dissolution of large rock fragments are also, in this case, larger than the "normal" primary pores. Recognition of "oversize" pores is important in the identification of secondary porosity. Photo by R. G. Loucks.

0.11 mm

Upper Cretaceous Teapot Sandstone
 Mbr., Mesaverde Fm.
Wyoming ca. 2,088 m (6,850 ft)

Slightly oversize pores and barely visible clay rims passing through pores are indicative of dissolution of unstable grains in this sample. The use of stained plastic or epoxy, introduced into the sample before it is cut or polished, is essential in order to differentiate true secondary porosity from grain removal during thin-section preparation.

0.10 mm

Upper Cretaceous Teapot Sandstone
Mbr., Mesaverde Fm. ★
Wyoming ca. 2,088 m (6,850 ft)

A large secondary pore (filled with blue plastic) produced by the selective leaching of several detrital grains. The secondary porosity is recognizable because of the presence of relict clay films which outline or cast former grains and because of the clearly "oversize" nature of the pores.

0.27 mm

Upper Cretaceous Tuscaloosa Fm. ★
Louisiana 6,230 m (20,439 ft)

A sample artificially etched for about 20 seconds in dilute hydrochloric acid. Removal of a detrital carbonate(?) fragment has left a thin, but apparently resistant, shell of authigenic chlorite cement. Such shells are sufficiently durable to survive diagenetic dissolution of core grains under subsurface conditions as seen in previous example.

SEM   40 μm

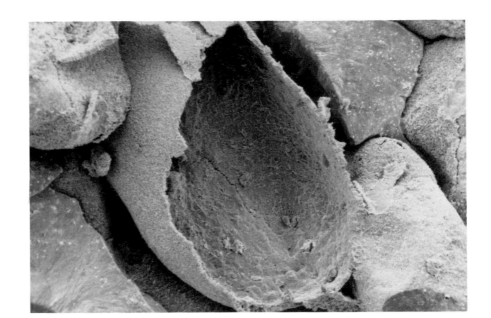

Upper Permian Bell Canyon Fm.
Texas 1,390 m (4,560 ft)

Corrosion or honeycombing of grains. In this example the feldspar grain in center has been partially leached along preferred crystallographic lines. The amount of remnant secondary porosity is small however, because of later overgrowth cementation on the feldspar. The recognition of such reduced secondary porosity is important in many reservoir rocks. Photo by C. R. Williamson.

SEM   10 μm

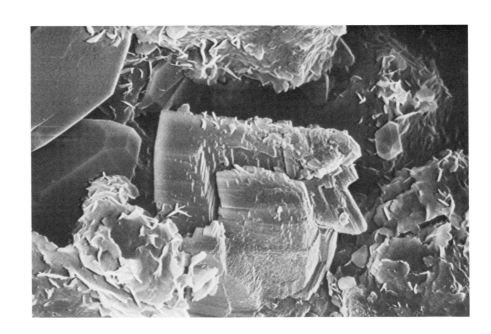

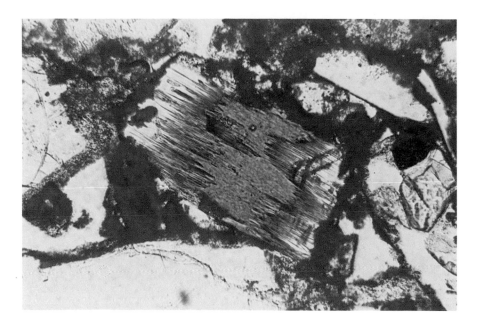

Miocene 'Hayner Ranch Fm.'
New Mexico

Although feldspars, carbonates, and evaporites are the most commonly leached minerals in sandstones, heavy minerals commonly are also affected. In this example, a detrital hornblende (center of photo) shows corroded margins with extremely thin terminations which clearly could not have survived transport. White area between greenish hornblende and red-brown authigenic hematite is pore space of secondary origin. Photo by T. R. Walker.

0.63 mm

Miocene 'Hayner Ranch Fm.'
New Mexico

SEM view of corroded, detrital hornblende from a sandstone. This detailed view of a grain similar to that shown in previous photo illustrates the very delicate etch features which can be developed during subsurface dissolution. The secondary porosity produced by such leaching becomes important for reservoir rocks only when it is interconnected by fractures, primary porosity, or elongate leached voids. Photo by T. R. Walker.

SEM   40 μm

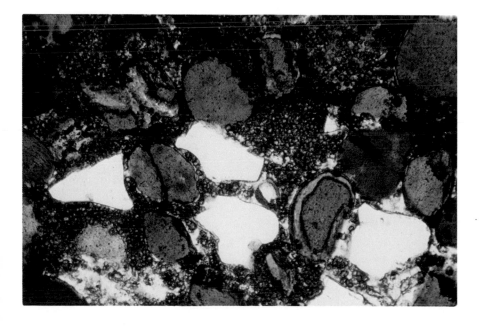

Upper Cretaceous Upper Logan
  Canyon Fm.
Canada (Scotian Shelf) 1,619 m (5,311 ft)

Secondary porosity produced by shrinkage of glauconite grains. Pore space shown in blue; glauconite is olive-green; quartz is white; small siderite crystals are brown. Shrinkage voids form through dehydration and/or recrystallization of minerals such as glauconite or hematite. Secondary porosity is preserved only where a sufficiently strong framework of cement exists to prevent compaction. Photo by D. A. McDonald.

0.27 mm

178

Upper Cretaceous Price River Fm.
Utah

Leaching of cements is an important factor in development of secondary porosity. In this example, calcite cement (stained red) has been partially dissolved under subsurface conditions. Note elongate porosity (blue) and slightly rounded margins of remnant calcite crystals. Subsurface leaching of carbonate minerals may result from meteoric flushing, biogenic CO_2 formation or maturation of organic matter in deeper subsurface environments.

0.06 mm

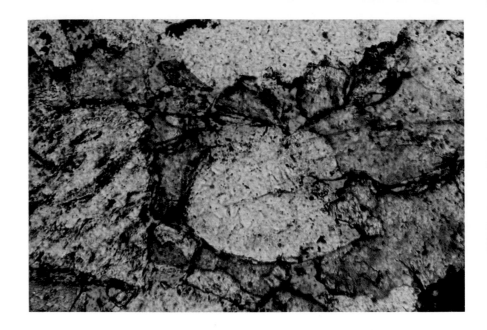

Lower Cretaceous sandstone
Canada (N.W.T.) 3,024 m (9,920 ft)

Another example of partial dissolution of calcite cement (stained red). Note preservation of corroded remnants of once apparently complete pore filling cement—this is an excellent criterion for the recognition of secondary porosity development. Quartz overgrowth cement preceeded calcite formation in this example, and was unaffected by the leaching of the calcite. Photo by V. Schmidt.

0.10 mm

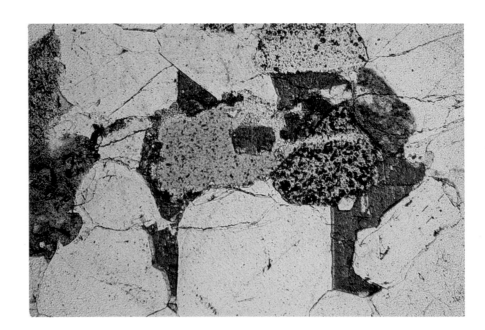

Upper Cretaceous Teapot Sandstone
 Mbr. of Mesaverde Fm.
Wyoming ca. 2,088 m (6,850 ft)

Leaching of carbonate cements can be quite patchy, as in this example, Calcite cement and detrital feldspars have been dissolved only in certain areas and are well preserved in other spots. Decementation patterns are commonly controlled by microfracture distribution or other fabric elements.

0.27 mm

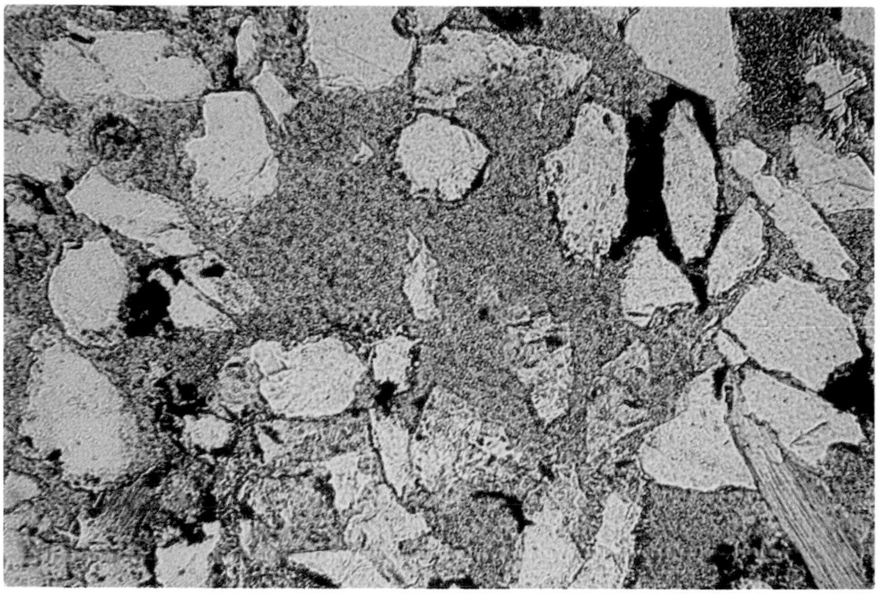

Jurassic sandstone (Statfjord field)
Norwegian North Sea 2,520 m (8,267 ft)

An extreme example of secondary porosity development. Nearly complete mesogenetic dissolution of sedimentary and eogenetic carbonate minerals created 40 percent porosity (by volume, based on core analysis). Such massive creation of secondary pore space in a rock which once was tightly cemented can have a major impact on reservoir distributions and properties. Photo by V. Schmidt.

0.27 mm

Cambrian Old Fort Island Ss.
Canada (N.W.T.)

Elongate secondary porosity. The elongate pores have resulted from relatively early-diagenetic marginal replacement of quartz grains by carbonate minerals. Subsequent dissolution of the carbonate minerals has left narrow but highly interconnected pore space which is a relatively rare but diagnostic characteristic of secondary leaching. Photo by D. A. McDonald.

0.10 mm

Upper Cretaceous Teapot Sandstone
Mbr., Mesaverde Fm.
Wyoming ca. 2,088 m (6,850 ft)

Microfracture porosity in a sandstone. Microfractures can be very effective in improving effective permeability of "tight" sandstones. It is difficult to prove petrographically whether fractures are original or were produced during coring or sample processing. Presence of oil stain on fracture walls or incipient cementation along fracture can be used as evidence of "original" fracturing.

0.10 mm

Upper Cretaceous Star Point Ss. *
Utah

Fracture porosity in a sandstone. The large, unfilled fracture (shown in blue) which transects sample provides a significant percentage of the total porosity in this sample and improves overall permeability as well. Oil staining along fracture indicates that fracture predated drilling and shows its importance to oil migration and production. Note small amounts of preserved primary intergranular porosity.

0.10 mm

Permian sandstone
Texas

Filled secondary (fracture) porosity. The fracture in this example has been completely healed by growth of calcite cement. The relative timing of fracturing, oil migration, and cementation can be of critical importance in reservoir geology.

XN   0.10 mm

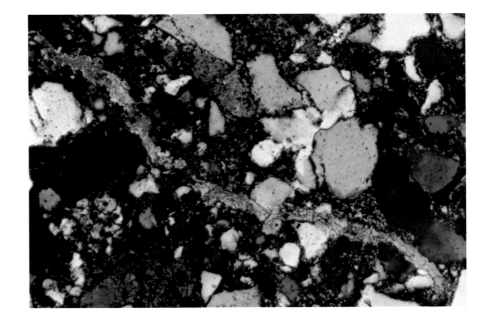

Lower Cretaceous sandstone
Canada (N.W.T.)

Fracture porosity can be created within individual grains. Here, a feldspar (photo center) has been fractured after partial leaching of carbonate cement and replacement crystals from the sample. Subsurface removal of cements commonly leads to fracturing of grains as the stabilizing and supporting material is removed and grain contacts begin to support a greater portion of the overburden stress. Photo by V. Schmidt.

0.25 mm

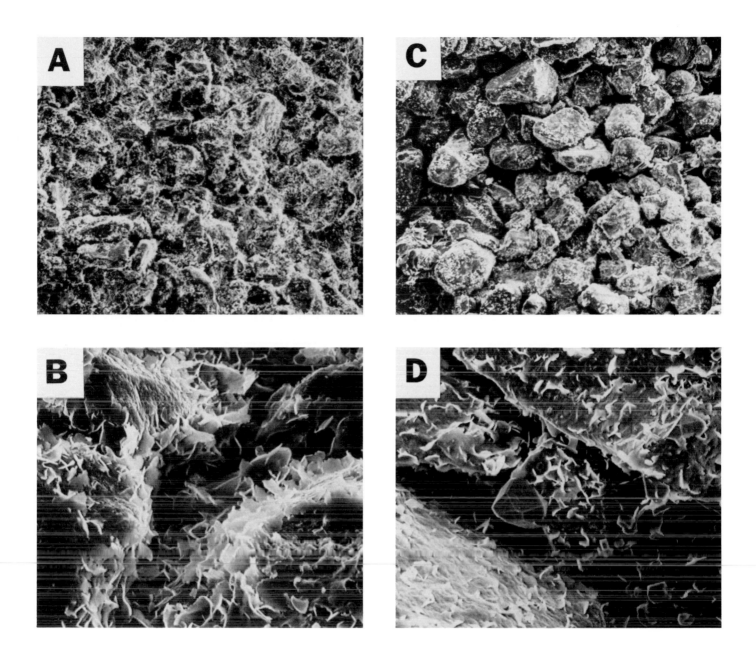

Significant information on porosity types can be obtained using scanning electron microscopy (SEM). Perhaps even more important, is that the SEM provides the ability to look at very small-scale textural relationships, pore-throat configurations, and other factors which affect sandstone permeability.

This series of photos illustrates the petrophysical-petrographic data obtainable with SEM. The sample in photos **A** and **B** is a high-porosity (21.2%), low-permeability (7.7 md) sandstone with abundant pore-lining chlorite. The sample in photos **C** and **D** is a high-porosity (26.3%), high-permeability (88.0 md) sandstone with some quartz overgrowth cement and only small amounts of authigenic chlorite. Thus, differences in permeability in sandstones with similar porosities can be related to differences in style and composition of cements.

All photos are from Permian Bell Canyon Formation, Texas; photos by C. R. Williamson. Scale is 110 μm for **A** and **C**; 11 μm for **B** and **D**.

Techniques

Techniques

Although petrography is an extremely valuable tool for the identification of minerals and their textural interrelations, it is best used (in many cases) in conjunction with other techniques.

Precise mineral determination commonly is aided by staining of thin sections or rock slabs, by X-ray diffraction analysis, or by microprobe examination. Minerals present in small amounts may best be analyzed after separation and concentration using heavy liquids, shaker tables, or other techniques. Likewise, noncarbonate minerals in a carbonate host rock are normally better analyzed in acid-insoluble residues than in thin section. Where detailed understanding of the trace element chemistry of the sediments is essential, X-ray fluorescence, microprobe, atomic absorption, or cathodoluminescence techniques may be applicable.

Commonly, sediments are too fine-grained for adequate examination with the light microscope. The practical limit of resolution of the best light microscopes is in the one to two micrometer (μm) range. Many detrital and authigenic grains such as clays, micritic carbonates, or organic matter fall within or below that size range. Furthermore, because most standard thin sections are about 30 μm thick, a researcher examines 10 to 20 of these small grains stacked on top of one another, with obvious loss of resolution. Smear mounts or grain mounts (slides with individual, disaggregated grains smeared or settled out onto the slide surface) are an aid in examining small grains where the material can be disaggregated into individual components. In most cases however, scanning and transmission electron microscopy are proved to be the most effective techniques for the detailed examination of fine-grained sediments.

Only a few of the most useful and petrographically oriented techniques will be discussed in this chapter. However, bibliographies at the ends of chapters throughout the book list references to techniques and applications relevant to specific minerals or diagenetic features. The bibliography at the end of this chapter provides additional references on general and specific analytical procedures. Although many of the techniques require sophisticated and expensive equipment, others, such as staining, acetate peels, insoluble residues, and grain separations can be performed in any laboratory. Scanning electron microscopes are now widespread in this country and virtually anyone can get access to one through a nearby university or company.

Because of the potential desirability of these techniques, it commonly is useful to prepare epoxy-cemented thin sections without coverslips. These sections can be examined under a light microscope either by placing a drop of water and a coverslip on the sample during viewing, or by using glycerine, mineral oil, or refractive index oils with (or without) coverslips. These methods may involve some loss of resolution, but do allow the cleaning and drying of the surface of the section and subsequent staining, cathodoluminescence, or microprobe examination. Uncovered thin sections also can be ground thinner than normal in cases where examination of very fine-grained sediments is needed. Finally, impregnation of sandstone samples with a darkly colored epoxy before thin section preparation will greatly facilitate the identification of the amounts and types of porosity present in the sample.

Clearly, one can spend a lifetime analyzing a single sample using all possible techniques. Efficient study requires a thorough understanding of all the available tools and proper application of the most useful and productive of these.

Staining

Staining techniques are among the fastest, simplest, and cheapest methods for getting reliable data on composition of detrital grains or cements in sandstones. Stains for calcite, organic matter, K-feldspar, and plagioclase are among the most useful stains.

The top photo shows a sandstone (Pennsylvanian Tensleep of Wyoming; 0.10 mm) which has been stained for calcite using Alizarin Red S. Calcite and dolomite are both present as cements but only the calcite has taken the red stain. Note imperfections in staining of calcite near thin edges of crystals and where bubbles were present. Because of the similarity in optical properties of calcite and dolomite, such staining is essential for accurate identification of these minerals.

The lower photo shows a K-feldspar (sanidine) from a Tertiary intrusive in Nevada (0.28 mm). This originally colorless grain was stained for potassium using a sodium cobaltinitrite solution. The lack of twinning and cleavage in this grain make it difficult to differentiate from quartz without time-consuming optical study. Staining, however, provides a rapid and reliable alternative for routine petrographic studies.

Directions for stain preparation are given in the references by Bailey and Stevens (1960), Laniz, et al (1964), Dickson (1966), Friedman (1971), and Whitlatch and Johnson (1974).

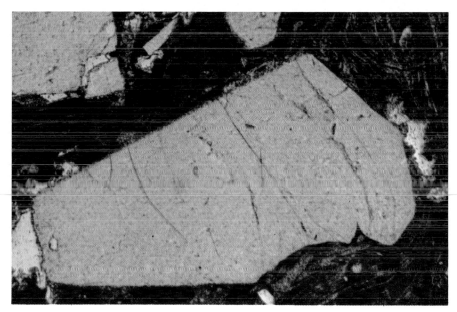

Cathodo-luminescence

Cathodoluminescence can be an invaluable tool in petrographic studies. It provides information on the spatial distribution of trace elements in terrigenous clastic (as well as carbonate) grains and cements. Analysis can be done using polished rock chips, polished thin sections, or even unpolished and uncovered thin sections. The equipment requires costs about the same as a moderately priced polarizing microscope and can be installed on virtually any microscope.

This example shows a sandstone from the Devonian Hoing Sandstone Member of the Cedar Valley limestone in Illinois. The upper photo, taken with transmitted light, shows a quartz arenite with elongate, sutured intergranular boundaries which might be considered as indicative of compaction and pressure solution. The lower photo, taken with cathodoluminescence, shows the same field of view with dramatically different results. The detrital grain cores, which luminesce orange and blue, can be seen to be well rounded, and touch each other only at point contacts. Subsequent quartz overgrowths (generally nonluminescent with some luminescent zones) have obliterated most porosity and give the appearance of a compacted fabric when cathodoluminescence is not used. The differences in luminescence between detrital grain cores and authigenic overgrowths is a function of subtle differences in their trace element composition.

These differences are accentuated by long-exposure-time photography of small areas because of the inherent weakness of the luminescence. Photos by R. F. Sippel.

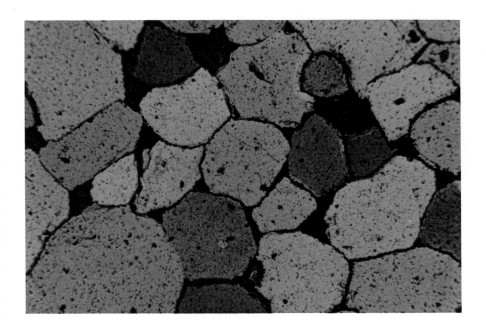

Fluid Inclusion Studies

Fluid inclusions can be found in virtually all crystals. They range in size from less than 1 μm to a few centimeters, although inclusions larger than 1 mm are uncommon. Most contain a solution which represents a sample of the original waters from which the crystal formed, plus a gas or solid phase which may have separated during cooling.

Careful petrographic study (commonly using heating or freezing stages) can determine the composition and original temperature of the fluids involved in crystal formation. This can provide useful information on the timing and conditions of cementation or mineralization, although care must be taken to determine the exact time relations of the fluid inclusions and the host mineral.

The top photo shows two phase (fluid and liquid) inclusions from the fluorite-zinc district of Illinois (0.40 mm).

The middle photo illustrates a three phase (solid, liquid, and gas) inclusion in a quartz geode from Iowa (10 μm).

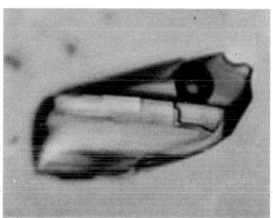

The bottom photo shows a two phase immiscible mixture of oil and water within a fluorite crystal from the same locality as the top photo. The colorless fluid is a strong brine; yellow fluid is oil; gas bubbles are methane associated with oil. All photos by Edwin Roedder.

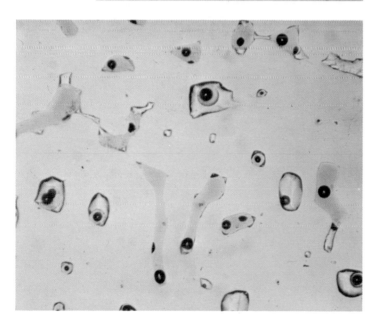

Scanning Electron Microscopy

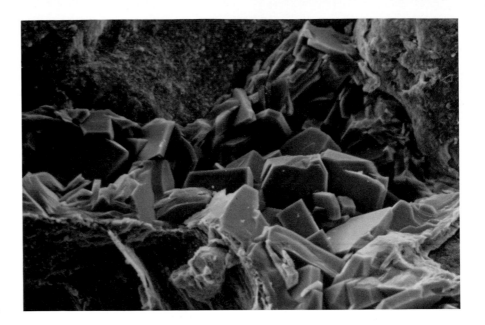

Scanning electron microscopy provides two major advantages over light microscopy: an extreme depth of focus and a wide range of magnification. The top photo, from the Miocene 'Hayner Ranch Formation' of New Mexico (30 μm), illustrates the remarkable depth of focus of the SEM. Clinoptilolite crystals here fill pore space in a sandstone.

The middle photo, from the Permian Rotliegendes Sandstone of the North Sea (7 μm), shows the excellent resolution of extremely small wispy terminations on authigenic illite cements.

The bottom photo, also from the Rotliegendes Sandstone, illustrates a specialized technique—pore casting. The rock was pressure-impregnated with epoxy and the component grains were leached out subsequently with hydrofluoric acid. This leaves a three-dimensional network of epoxy which shows the geometry of the pore system, including small but interconnected pores not normally seen in thin section.

Top photo by C. W. Keighin (courtesy of T. R. Walker); lower photos by E. D. Pittman.

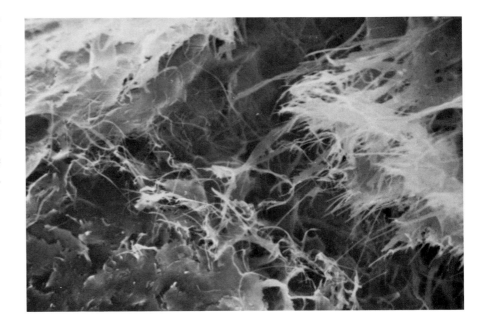

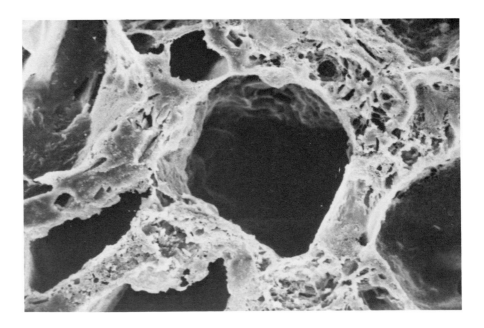

Energy Dispersive Analysis

Scanning electron microscopy is useful not only for examining sediment textures, but, when equipped with an energy dispersive analyser, it can be used effectively for mineral identification and semi-quantitative chemical analysis. The analyses are rapid (seconds) and require relatively little sample preparation. In most cases, small chips of the sample can be mounted on a small plug with no polishing or cutting required. The sample is then coated with a gold-palladium alloy (or other conductive metal) and is inserted into the SEM.

Here, a potassium feldspar from the Cretaceous Frontier Formation of Wyoming is shown with an accompanying analytical spectrum. Although the grain might be identifiable as a feldspar on the basis of its crystal shape, cleavage, and other features alone, the energy dispersive analysis provides additional chemical data which allows positive identification.

The analytical trace (lower photo) shows major peaks for Si and K (the main K peak has the long, pale blue line over it) with only very minor peaks for other elements. This indicates a rather pure K-feldspar composition.

Although energy dispersive analysis on the SEM provides an excellent tool for mineral identification, it is not ideally suited for quantitative analytical work. Detailed determination of mineral composition or analysis of trace element contents of small crystals is best done using polished samples on an electron microprobe.

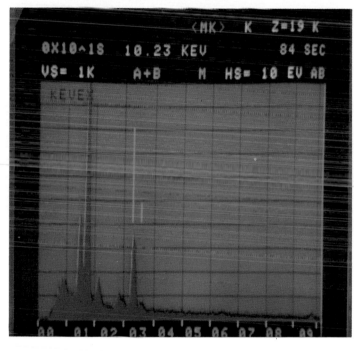

Electron Microprobe Analysis

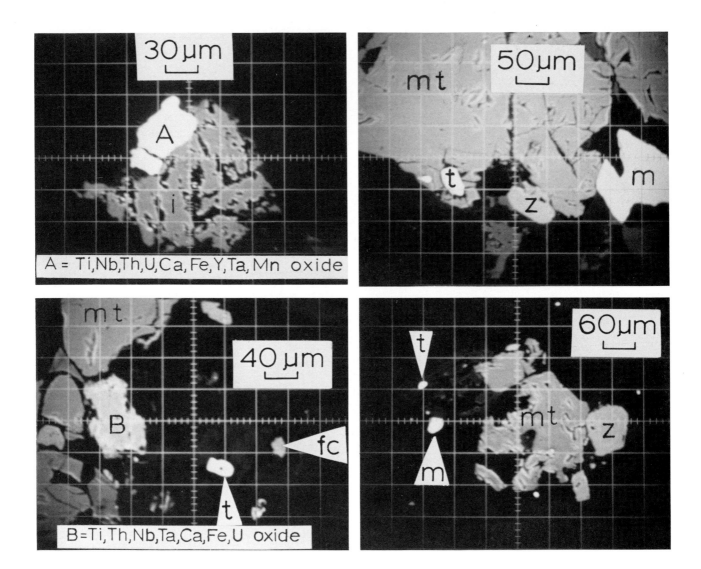

The electron microprobe, used in the sample current-image mode, is useful for study of textural relations of minerals. Contrast in the images produced is due to high-versus-low atomic number. On an electron microprobe equipped with a multichannel analyzer/energy dispersive detector system, mineral grains can be chemically characterized in a few seconds according to their spectra of elements.

The four figures are electron microprobe sample current image photographs of granite in which the heavy minerals magnetite (mt), ilmenorutile (i), zircon (z), monazite (m), thorite (t), and fluocerite (fc) are surrounded by quartz and potassium feldspar (black).

Microprobe sample preparation (polished sections) is relatively time consuming, and analytical work using the microprobe is relatively complex and expensive. However, in many cases, microprobe analysis provides the only reliable method for obtaining quantitative data on major and minor element composition of minerals or accurate identification of extremely small grains. Photos by G. A. Desborough.

Selected Techniques Bibliography

Anderson, C. A., 1973, Microprobe analysis: New York, John Wiley and Sons, 571 p.

Bailey, E. H., and R. E. Stevens, 1960, Selective staining of K-feldspar and plagioclase on rock slabs and thin section: Am. Mineralogist, v. 45, p. 1020-1025.

Borg, I. Y., and D. K. Smith, 1969, Calculated X-ray powder patterns for silicate minerals: Geol. Soc. America Memoir 122, 896 p.

Carroll, Dorothy, 1970, Clay minerals: a guide to their X-ray identification: Geol. Soc. America Spec. Paper 126, 80 p.

Carver, R. E., ed., 1971, Procedures in sedimentary petrology: New York, Wiley-Interscience, 672 p.

Cohen, A. D., and W. Spackman, 1972, Methods in peat petrology and their application to reconstruction of paleoenvironments: Geol. Soc. America Bull., v. 83, p. 129-142.

Dickson, J. A. D., 1966, Carbonate identification and genesis as revealed by staining: Jour. Sed. Petrology, v. 36, p. 491-505.

El-Hinnawi, E. E., 1966, Methods in chemical and mineral microscopy: New York, Elsevier, 222 p.

Friedman, G. M., 1971, Staining, in R. E. Carver, ed., Procedures in sedimentary petrology: New York, John Wiley and Sons, p. 511-530.

Goldstein, J. I., and Harvey Yakowitz, eds., 1975, Practical scanning electron microscopy; electron and microprobe analysis: New York, Plenum Press, 582 p.

Gorz, H., R. J. R. S. B. Bhalla, and E. W. White, 1970, Detailed catho-doluminescence characterization of common silicates: Penn. State Univ., MRL Spec. Pub. 70-101, p. 62-70.

Kicke, E. M., and D. J. Hartmann, 1973, Scanning electron microscope application to formation evaluation: Gulf Coast Assoc. Geol. Socs., Trans., v. 23, p. 60-67.

Laniz, R. V., R. E. Stevens, and M. B. Norman, 1964, Staining of plagioclase feldspar and other minerals with F. D. and C. Red No 2: U.S. Geol. Survey Prof. Paper 501-B, p. 152-153.

Leach, D. L., R. C. Nelson, and D. Williams, 1975, Fluid inclusion studies in the northern Arkansas zinc district: Econ. Geol., v. 70, p. 1084-1091.

Long, J. V. P., and S. O. Agrell, 1965, The cathodo-luminescence of minerals in thin section: Mineralogical Mag., v. 34, p. 318-326.

Marshall, D. J., 1978, Suggested standards for the reporting of catho-doluminescence results: Jour. Sed. Petrology, v. 48, in press.

Müller, German, 1967, Methods in sedimentary petrology: New York, Hafner, 283 p.

Pittman, E. D., and R. W. Duschatko, 1970, Use of pore casts and scanning electron microscope to study pore geometry: Jour. Sed. Petrology, v. 40, p. 1153-1157.

Potosky, R. A., 1970, Application of cathodoluminescence to petrographic studies: The Compass, v. 47, p. 63-69.

Price, Ilfryn, 1975, Acetate peel techniques applied to cherts: Jour. Sed. Petrology, v. 45, p. 215-216.

Roedder, Edwin, 1971, Fluid-inclusion evidence on the environment of formation of mineral deposits of the southern Appalachian Valley: Econ. Geol., v. 66, p. 777-791.

——— 1972, Data of geochemistry; composition of fluid inclusions: U.S. Geol. Survey Prof. Paper 440 JJ, p. JJ1-164.

Sippel, R. F., 1965, Simple device for luminescence petrography: Review Sci. Instruments, v. 36, p. 1556-1558.

Smith, J. V., and P. H. Ribbe, 1966, X-ray emission microanalysis of rock-forming minerals. III. Alkali feldspars: Jour. Geology, v. 74, p. 197-216.

——— and R. C. Stenstrom, 1965, Electron-excited luminescence as a petrologic tool: Jour. Geology, v. 73, p. 627-635.

Spencer, Charles W., 1960, Method for mounting silt-size heavy minerals for identification by liquid immersion: Jour. Sed. Petrology, v. 30, p. 498-500.

Sweatman, T. R., and J. V. P. Long, 1969, Quantitative electron microanalysis of rock-forming minerals: Jour. Petrology, v. 10, p. 332-379.

Tickell, F. G., 1965, The techniques of sedimentary mineralogy: Developments in sedimentology, Volume 4: Amsterdam, Elsevier, 220 p.

Weinbrandt, R. M., and Irving Fatt, 1969, A scanning electron microscope study of the pore structure of sandstone: Jour. Petroleum Tech., v. 21, p. 543-548.

Whitlatch, H. B., and R. G. Johnson, 1974, Methods for staining organic matter in marine sediments: Jour. Sed. Petrology, v. 44, p. 1310-1312.

Explanation of Indexing

A reference is indexed according to its important, or "key" words.

Three columns are to the left of the keyword entries. The first column, a letter entry, represents the AAPG book series from which the reference originated. In this case, M stands for Memoir Series. Every five years, AAPG merges all its indexes together, and the letter M will differentiate this reference from those of the AAPG Studies in Geology Series (S) or from the AAPG Bulletin (B).

The following number is the series number. In this case, 28 represents a reference from Memoir 28.

The last column entry is the page number in this volume where the reference will be found.

Note: This index is set up for single-line entry. Where entries exceed one line of type, the line is terminated. (This is especially evident with manuscript titles, which tend to be long and descriptive.) The reader sometimes must be able to realize keywords, although commonly taken out of context.

Index

198